EVERYDAY LIVES IN THE
GLOBAL CITY

Based on ethnographic research in London, this study investigates globalization as a feature of everyday life. Rejecting simplifying notions of globalization as a macro-economic force, it provides a grounded picture of various ways in which people's biographical trajectories are tied up with the global cultural economy. While recent debates in the social sciences started to investigate the link between globalization and biography in general, this study more specifically stresses the catalytic function of the global city environment.

The idea of 'milieu' has always linked people to their surroundings. The main argument developed throughout the book is that the gloablization of their lives is experienced by people as the 'extension' of their milieux, both spatially and symbolically. This book shows that within a global cultural economy people's milieux can potentially span the world. Grounded in life histories and local history, it develops a concept of milieu that emphasizes the symbolisms of belonging against the primacy of place.

Jörg Dürrschmidt is Lecturer in Sociology at the University of the West of England.

EVERYDAY LIVES IN THE GLOBAL CITY

The Delinking of Locale and Milieu

Jörg Dürrschmidt

Routledge
Taylor & Francis Group

LONDON AND NEW YORK

First published 2000
by Routledge
2 Park Square, Milton Park, Abingdon, Oxfordshire OX14 4RN

Simultaneously published in the USA and Canada
by Routledge
711 Third Avenue, New York, NY 10017

First issued in paperback 2015

Routledge is an imprint of the Taylor and Francis Group, an informa business

© 2000 Jörg Dürrschmidt

Typeset in Perpetua by Keystroke, Jacaranda Lodge, Wolverhampton

British Library Cataloguing in Publication Data
A catalogue record for this book is available from the British Library

Library of Congress Cataloguing in Publication Data
Durrschmidt, Jorg, 1964–
Everyday lives in the global city : the delinking of locale and milieu / Jorg Durrschmidt.
p. cm.
1. Sociology, Urban—England—London. 2. City and town life—England—London.
3. London (England)—Social conditions. I. Title.

HN398.L5 D87 2000
307.76′09421—dc21
00–030821

ISBN13: 978-1-138-86729-1 (pbk)
ISBN13: 978-1-84142-011-0 (hbk)

CONTENTS

Acknowledgements   viii

Introduction   1

1   **The (global) city as a *pars pro toto* of (global) society?**   4

The legacy of Socrates 4
Metropolis and modern civilization 5
Urbanization versus globalization? 8
The global city and its microglobalized hinterland 12

2   **Towards a phenomenology of globalization**   16

A phenomenology of everyday life 16
A phenomenology of extended milieux 18
Re-locating the concept 22

3   **Eight London biographies – lives in the globalized
    world city**   25

Sarah Penhaligan – the 'mobile cosmopolitan' 27
Ulla Anderson – the 'settled cosmopolitan' 28
Harold Ford – the 'global business man' 30
Nicos Calacuri – the 'local entrepreneur' 32
Herbert Reading – the 'retired local' 34
Ira Braithwaite – the 'metropolitan' 35
Barbara van der Velde – the 'expatriate' 37
Rolf Grauer – the 'reluctant explorer' 39

CONTENTS

4 The uprooting of milieux   **42**

External uprooting – coming out 43
Internal uprooting – leaving the estate 48
Disrupting the local milieu 54

5 The delinking of locale and milieu   **60**

Mobile and generalized milieux 61
The extended milieu – between 'here' and 'there' 68
The convergence of 'here' and 'there' 74
'Home' as a significant place 77
Environments of like-mindedness – St Anne and St Agnes Church 81

6 Streatham – the reluctant suburb: the metropolis expands   **91**

Streatham – the would-be West End of South London 93
Streatham High Road – early links between Streatham and London 98
Dr Johnson's Streatham – early stratification of local and extended milieux 101
The metropolis absorbs Streatham 106
Struggle for local identity – the disruption of a local milieu 108

7 Extended milieu and 'soft city' – generating symbolic space   **115**

'Soft city' and 'concept city' 116
Symbolic space and the concept of the city 117
The 'soft city' as extended milieu 123
Milieu and metropolitan variety 127

8 The globalized world city and its 'cosmion'   **132**

The globalized cosmion 132
Moving in sociospheres – the 'unknown others' 140
A city of 'neighbours'? 153
The 'distant' neighbour 157
Neighbourhood community turned socioscape 161

Conclusion – the end of the world city? 172

Notes   175
Bibliography   178
Index   183

ACKNOWLEDGMENTS

Writing is by definition a rather lonesome process. However, there are many people who, in various ways, have helped me in producing this book. Not all can be mentioned here. I would like to thank Richard Grathoff and Rüdiger Korff, at Bielefeld University, for their initial inspiration concerning the Phenomenology of Global Cities. Many thanks to Ruth and Ron Englund for providing a 'home from home' during my field work in London, as well as for the shared enthusiasm for Crystal Palace FC. Intellectually I have greatly benefited from my attachment to the 'Globalization Research Cluster' at the Roehampton Institute, London, during various periods of field work. I would like to thank everyone there for many good discussions and the intellectually stimulating atmosphere. I am particularly indebted to Martin Albrow, currently at the Woodrow Wilson Center, Washington, for his intellectual enthusiasm and persistent encouragement throughout this book project. Many thanks to Neil Washbourne, now at Leeds Metropolitan University, for engaging discussions on microglobalization, and the many days spent just chatting and wandering around London's second hand book shops. Amongst my colleagues at the University of the West of England I would want to acknowledge the influence of Pete Jowers with his scepticism towards all big paradigms, as well as the pleasure I gained from teaching and researching around issues of global transformation with Bill Hill and Graham Taylor. For effective and patient guidance through the editorial and technical side of this book project, I would like to thank Caroline Wintersgill, Mari Shullaw and James McNally, as well as the Keystroke team. Finally, I would like to thank Sue, who was a good companion during the good and not so good times involved in writing this book. But most importantly, thanks to those individuals who agreed to talk to me for hours about their lives, London, and the wider world. I hope their voices can still be heard in this book.

INTRODUCTION

This book could well be introduced as yet another 'thick description' of everyday life in London, the world city. However, the book does not aim to provide a comprehensive picture of metropolitan life. Instead, it provides a descriptive analysis of how people who live in the global city are participating, to different degrees and in various ways, in the emerging global cultural economy. The main concern pursued throughout the book is to observe how this involvement affects people's milieux. That is, the familiar spatial and social environment in which they maintain their daily routines. In this respect it shall then be argued that these milieux 'extend' not only beyond immediate local surroundings, but beyond the metropolis as such, thereby in turn transforming the very make up of London's everyday life.

The adventure of this book then is to investigate the intersection of two complementary processes: the microglobalization of the world city's everyday life, and the globalization of the biographies that are participating in it. To achieve this, the book links narrative accounts of London's everyday life with the contemporary debates on globalization and individualization.

While the biographies of eight Londoners will provide 'prisms' through which globalization processes in London's everyday life are being reflected, London itself will be described as a 'base' from which these biographies (have to) engage with global society, both in and beyond the global city.

I regard this book as a first step towards a phenomenology of globalization. While the contemporary debate still predominantly perceives of globalization as a macro problem of economics, finance, politics, and culture, my argument describes and analyses globalization processes as a lived problem in the world of everyday life. Or more precisely, everyday life in a global city. Again, while most of the debate on global or world cities is concerned with their status as key nodes of a global cultural economy, this book looks at London as an intersection of globalized life worlds. 'Everyday Lives' in the title refers to the phenomenological perspective pursued in this book. It attempts to reveal some of the everydayness of globalization within the milieux of people based in London, but by no means claims to take in the entire

1

complexity of London's everyday life. The plural s in the title is intentionally chosen in order to reflect the relatively detached and unrelated coexistence of different social and personal worlds within the global city.

While most of the ethnographic data in this book are biographical, the actual argument of this study is about space: the globalization of urban space, and the spatial restructuring of milieux interwoven within it. This is where the book hopes to make a genuine contribution to what Roland Robertson demands from social sciences in response to 'globaloney': the exploration of concrete forms of spatial and temporal structuring of the world in which we live. The argument developed invites the reader to explore the idea that in the context of globalization and individualization processes the individual's milieu is one such form in which a transient and ephemeral global environment is being (re)appropriated and familiarized. Especially in its 'extended' form – that is, stretching across significant places and not confined to a locale – the milieu provides a crucial medium for 'time–space distanciation', characterized by the intersection of presence and absence, and of 'here' and 'there', subsequently leading to new forms of situatedness beyond a nostalgic understanding of 'home'. This investigation of globalization processes, as experienced in people's biographies or milieux, places the argument of this study outside the conventional dichotomies of global/local and micro/macro.

As the emphasis on a biographical approach might have suggested to the reader already, this is a study committed to 'grounded theory', open to the challenges everyday life throws at theoretical assumptions, and gaining inspiration for the modification of analytical concepts from people's own interpretation about the world around them.

At the outset of my study I had no idea I would eventually write about micro-globalization in everyday life, and about extended milieux. What I initially intended was a study on contemporary metropolitan milieux, in its approach located in the tradition of Simmel and the Chicago School, but inspired by the more recent writings of Bahrdt, Sennett and Hannerz, while using the phenomenological concept of 'milieu' in the tradition of Scheler, Schütz, and Goffman as my analytical tool.

Accordingly, I had a fairly general interest in how people generate familiarity and situatedness in a huge place like London; how, in other words, they would generate their respective 'soft cities' out of the complex and changing urban organism called London. What I had in mind for the first couple of interviews as the typical challenge for a metropolitan milieu, were the Simmelian ideas of disorientation and 'sensory overload'. I was then surprised to see these problems marginalized as temporary challenges for the newcomer. Accordingly, these issues have been confined to one chapter. The real challenge articulated by the interviewees was, on the one hand, the problematic character of local living, affected by individual mobility and the resulting transitoriness of social relationships, and, on the other hand, the importance of keeping in touch with places and people across the world. This should perhaps not have come as a surprise, as today's London hardly has an 'indigenous' population, and as most Londoners are Londoners 'by adoption'.

What my interviewees presented me with was the lived reality of what could perhaps best be described as an 'extended milieu'. What this term tries to capture is the observation that the milieu, defined as the individual's familiar field of action and experience, can no longer automatically find situatedness in the constancy of the social and material environment of the locale. Instead people maintain their milieux by engaging, not always by choice, with global society, both within and beyond the metropolis. It goes without saying that this notion of the 'extended milieu' does not just imply a geographical stretching, but has also a social and symbolic dimension. The idea of an 'extended milieu' was then further explored in another sample of semi-structured interviews, out of which eight were selected for a thorough analysis.

Out of this crystallized the main thesis of this book: the delinking of milieux from locales, and their subsequent extension. The different chapters of the book follow different dimensions of this process, from the 'uprooting' of the milieu to the development of a 'symbolic territory of the self'. While the argument is mainly grounded in the narratives of eight 'Londoners', the argument finds complementary illustration in historiographic sketches. In particular, chapter 6 looks at the delinking of locale and milieu from the perspective of a particular London locality.

The nature of 'grounded theory' is that it could go on and on. As such the aspects and dimensions of 'extended milieux' based in the global city's everyday life as explored in this book, are by no means final. Each of these dimensions could be further analysed, specified, other dimensions could easily be added. I therefore consider this book to be a first step in a certain direction which needs further pursuing. If the reader should find that it raises more questions than it answers, the author would see this as an achievement.

1

THE (GLOBAL) CITY AS A
PARS PRO TOTO OF
(GLOBAL) SOCIETY?

I am you see, a lover of learning. Now the people in the street have something to teach me, but the fields and trees won't teach me anything.

Socrates (cf. Plato 1977: 26)

The legacy of Socrates

'Phaedrus', Socrates' partner in the famous dialogue of the same name, is amazed by Socrates' proud acknowledgement that he is a citizen of Athens, but only a 'visitor' to its adjoining countryside, and although he promptly accuses him of living a life confined within the walls of the city, Socrates is not touched by this criticism. The city has plenty to offer for the philosophically minded. Moreover, it is *the* place to be for the one who is interested in the miracles of human nature (Plato 1977: 26).

In a metaphorical sense, Socrates' stand towards the city as a distinguished setting for understanding human society reflects the situation in current social sciences. He has given up on the 'grand narratives', in his words: 'rationalizing legends', and returned to basic studies of the social at hand. He intends to follow the Delphic injunction 'to know yourself', not through abstract philosophizing, but as a task in daily life, listening to and discussing with the plethora of differing and contradicting local voices. What better place is there than the city to do that?

And indeed, looking at the dominance of urban-based studies in the context of different attempts to grasp a contemporary society in transition, 'Exploring the City'[1] seems to be the thing to do for the contemporary social scientist hoping to grasp at least some features of a social reality increasingly escaping established theories and concepts. Does this mean then that understanding the city is the same as understanding society?

4

There is certainly an implicit assumption in this direction in Socrates' speech about Athens. Mumford raises our attention to the 'cockney illusion' in Socrates' thinking about the city and human society. According to him, it is Socrates' static image of city culture that does neither justice to human nature nor to the city's embeddedness in wider society. What Mumford re-emphasizes on the one hand is the dependency of the city on its surroundings, important not only for its natural metabolism, but 'equally nourishing to the mind'. In other words, the uniqueness of city culture is largely due to the active engagement with its hinterland. On the other hand, Mumford stresses as one of the most important features of human nature 'that of transcending natural limitations' (1991: 199ff.).

So while this book follows Socrates' assumption that perhaps in the everyday-ness of city life and in engagement with the city dweller we can learn more about human nature and human society than in any other place, it also follows Mumford's argument which emphasizes the structural link between the uniqueness of city culture on the one hand, and its multi-layered dependency on its surroundings on the other. In other words, while the urban dweller's life might serve as a 'window' through which to investigate tendencies in society at large, it does so because city life properly understood, will imply expanding our horizon beyond the actual walls of the city.

While Socrates might be the first one to have articulated the city as a distinguished site of social research, the idea that the city could be seen, admittedly in very different ways, as a *pars pro toto* of society, has kept its prominence in social science ever since, with a particular resurgence in modern age. The following section will provide a selected review of sociological ideas within the context of modernity that link the city and urban culture to developments in wider society. This short review should serve to set the stage for the question that inspires the argument developed in this book: what is the link between the global city's everyday life and global society?

Metropolis and modern civilization

It is impossible to start this review without mentioning a philosopher who is not exactly modern and yet has influenced modern thinking with his ideas on the city perhaps more than anyone else. Aristotle's concept of the Greek *polis*, as outlined in *The Politics*, is relevant here not so much in its descriptive elements regarding the 'good life' to be gained in the polis, but with regard to the structural issues it raises. There is first the issue of the size of the city, insofar as for Aristotle urban culture can only be equated with the 'good life' for a limited number of people for whom social interaction is based on mutual acquaintance. Second, there is the observation that the spatial form of the city supports a certain social order. As Mumford puts it, 'his true innovation consisted in realizing that the form of the city was the form of its social order, and that to remould one it is necessary to introduce appropriate changes in the other' (1991: 202, 216ff.). Now, these are certainly two issues that still figure prominently in the debate on modern and global cities respectively. Can their status be defined by size, and to what extent does their built environment reflect the social order at the centre of which these cities supposedly are?

In modern age, then, we find a generation of philosophers and sociologists explicitly relating the city's visible development into a dominant driving force of industrialization and urbanization, to the invisible destiny of a whole civilization. It was Spengler in his work *The Decline of the West* (1971) who coined the well-known phrase 'world history is city history'. The city in Spengler's account symbolizes (wo)man's status passage from a 'soil-rooted peasant' to a 'civilized town-tied intellectual nomad'. For Spengler the 'world city' is not only the place where 'the course of world history ends by winding itself up', but the 'city-as-world' also governs its surroundings via means of communication, transmitting the city's spirit into its surroundings, and making it more and more like it (1971: 89ff.). So, here in Spengler we have an early understanding of the dialectical process characterizing the world city's position in modernity: mirroring and concentrating the development of an ever-enlarging modern world, and at the same time increasingly stamping an urban culture onto this modern world.

Weber (1966) and Simmel (1969) have subsequently taken up some of Spengler's ideas in a more sociological approach. Weber's interest in the city, much like his interest in the world religions, has to be seen in the context of his attempt to explain the development of Western rationality, or more precisely, the unique role of rationalization in the development of Western European culture. Accordingly, Weber's interest is aimed at a medieval city that not only provides an enclave of freedom from the constraints of feudal society ('Stadtluft macht frei' / 'city air makes free'), but also fosters the conditions for rational conduct in all spheres of life. For Weber this is embodied in the development of entrepreneurship, guilds, and the city's civic institutions. Once again, the interesting point in Weber's argument is that he does not attempt to explain the medieval city as such, but considers its relevance for modern civilization. What is of interest to us here is the fact that during a short period when the city was truly a world in itself, principles and institutions of rationality developed that, one could claim, still govern the world (1966: 65f., 94ff., 181ff.).

For Simmel, understanding the 'modern age' means understanding the 'metropolis'. The metropolis provides the stage on which the 'dialectical dualism' between 'individual culture' and 'objective culture', the 'universal formula' of modernity, is played out in the extreme. The metropolis, according to Simmel, is the impersonalized crystallization of all that modernity has developed. In this context the individual can, on the one hand, find the stimulation and challenges needed to develop as a free and unique individual. On the other hand, it is equally possible that the individual should find it difficult amidst the abundance of commodified culture to gain and maintain unique individuality (1969: 385ff., 409ff.). It is certainly no overinterpretation to say that here we can see an early formulation of the 'individualization thesis', recently brought back into the globalization debate by Beck (1992: 127ff.) and Melucci (1996). Reading Simmel's account on the metropolis from this perspective, his suggestion that the urbanite is continuously engaged in the 'working out of a way of life', is possibly more revealing than the well-known statement concerning the need for adaptation to an 'intensification of nervous stimulation' (1969: 410, 420). What is also relevant for us in Simmel's argument is the reference to the metropolis not as a closed social unit, but as a centre of power that extends its influence beyond its immediate confines, not least through the universal symbol of money (1969: 411ff.). Freely summarizing Simmel's argument, it is in the metropolis that the overall tendencies of modernity come to the fore more clearly, so that it can justifiably be regarded as the *pars pro toto* of modern society.

The Chicago School, influenced by Spengler and Simmel, follows this idea to such an extent that for its major figures, Park and Wirth, modern sociology equates to urban sociology (Lindner 1990: 94ff.). For them, understanding Chicago's changing social organism was a way to grasp American society as an immigrant society (Smith 1988: 2ff.). Within this context, Park is more interested in the internal ecology of the city – its spatial structuration – following the ethnic, social and occupational segregation of its population (Park et al.1968: 1ff.). Wirth, in turn, following Simmel's ideas, is concerned with 'urbanism as a way of life', which, though typically a function of population size, density and heterogeneity most likely to be found in the actual city, is, however, not confined to the city walls. It is, according to Wirth, the fact that Western civilization readily embraces urbanism as its cultural form, or 'mode of life', that makes it 'distinctively modern'. Of interest in Wirth's argument is also the association of urbanism with transient social relationships, and the ephemeral character of places and people (Wirth 1969: 60ff.), as today this association is readily made with life in the global city.

Finally, Wirth offers us one of the most eloquent definitions of the world city or metropolis as the 'initiating and controlling centre of economic, political, and cultural life that has drawn the most remote communities of the world into its orbit and woven diverse areas, peoples, and activities into a cosmos', and subsequently, 'has brought together people from the ends of the earth' (1969: 61, 69). What Wirth describes here is something of an enlarged polis, the centre of gravity of a social cosmos, with things and people going 'its' way. For a time Chicago was certainly such a centre of gravity in this way, pulling resources and people into the American dream. But, perhaps, Wirth's definition fits even better those cities that Braudel describes as the command and control centres at the heart of regional world economies or empires (cf. Jones 1990: 45ff.). In this regard, at the turn of the century, London was the ultimate world city or metropolis, governing an empire that was a world in itself. Strolling through London, or visiting the 1851 Great Exhibition at the Crystal Palace, would by then tell us as much about the British Empire as any attempt to travel it.

This view of the metropolis as a centre of gravity of a large social universe – which comes through in all the snap shots of social theory on modernity and the city that we have taken in this section – has started to scramble over the last three decades. As Knight and Gappert (1989: 15ff.) argue with reference to the old ancient idea of the polis, the metropolis starts to lose control of its own development within the context of globalization. Thus, 'We no longer think of the metropolis as an autonomous self-governing polity but rather as an urban agglomeration that is shaped by forces beyond its control.'

As always in times of transition and paradigm shifts, this insight did not come suddenly. Mingione (1986: 145f., 149), summarizing the debates in urban sociology during the 1970s, talks about a 'long-term theoretical crisis' of urban studies that started, but was by no means overcome then. In the context of a much more complex debate it became clear that the 'urban' in itself is no longer a convincing explanatory unit, and that a mode of theorizing that carries urban phenomena 'beyond the confines of simply urban topics' was needed. Although Mingione already talks about a new complexity in society due to the 'computer-age revolution' and the new interplay between 'local, national, and super-national needs', the new mode of theorizing is not yet explicitly associated with globalization. Instead, the new approach was initiated with Castells (1977), who raised 'The Urban Question' in the context of a structuralist analysis of capitalism, firmly allocating the city its place within the reproduction of capitalism, namely as the spatial unit of collective consumption. From this perspective it should later become clear that capitalism itself underwent a transition, which would re-emphasize major cities as centres of, this time, global control and initiation.

Urbanization versus globalization?

If one is to approach globalization initially in quite general terms as 'the process whereby the population of the world is increasingly bonded into a single society' (Albrow 1993: 248f.), then urbanization can certainly be regarded as an integral part of it, if not *the* driving force behind it. It is, therefore, not surprising to find the transformation of late modern society ascribed to the continuation of urbanization, which was initially responsible for the shaping of early modern society.

Mumford, for example, captured the zeitgeist of the 1960s, when, in his monumental work *The City in History*, he stresses the link between Aristotle's polis and the urbanized modern world, claiming:

> This book opens with a city that was, symbolically, a world: it closes with
> a world that has become, in many practical aspects, a city.
>
> (1991: preface)

For the 1970s Lefèbvre announced the 'Urban Revolution', resulting in the 'complete urbanization of society' (1990: 7). And, as recently as 1990, Gottman re-emphasized a perspective that views recent changes in society 'largely as a result of continuing rapid urbanization' (Gottman and Harper 1990: 1). Finally, and most relevant to the approach taken in this book, Berger et al. (1974) explicitly link the breaking up of 'home worlds' and the subsequent 'pluralization of social life-worlds' to the 'urbanization of consciousness'. For them the city and its intrinsic sociocultural pluralism is one of the 'carriers' of modern consciousness and standards in society (1974: 15f., 64ff.).

One could now argue that these approaches, which stress a continuing process of urbanization throughout modernity, are compatible with that strand in globalization theory that stresses the continuity between modernization and globalization, and consequently regards the multi-layered dimensions of globalization as 'consequences of modernity'. From this perspective, especially in the aftermath of Berger's et al. statement about the city as 'carrier' of modernity, it is indeed rather surprising to see the city/urbanization not mentioned as one of the 'institutional dimensions of modernity', and respectively, the system of world cities as one of the 'dimensions of globalization', in Giddens' institutional model of globalization (Giddens 1994: 59, 71). Surely, staying within the context of Giddens' argument, the process of urbanization contributes to the universalization of the modes of life that initially developed in (Western) European modernity, and the (world) city is a place where the 'consequences of modernity' become radicalized and more visible?

However, today it is obvious that globalization is more than the incorporation of an increasing number of the world's population into cities, or the worldwide dominance of an urban way of life, not to mention regional differentiation within this urban way of life, reflected in the debates on 'de-urbanization' and 'over-urbanization' (Lefèbvre 1997: 78; Smith 1996: 13f., 188ff.; Savage and Warde 1996: 314f.). And, as Mingione (1986: 149) carefully articulates with regard to the long-standing theoretical crisis within urban studies, there was, however reluctant, a sense of realization that the various articulations of the urban in modern society had to be seen in the context of overall changes transforming contemporary society.

This sense of a new configuration of society beyond modernity is possibly most forcefully expressed in the concept of a 'Global Cultural Economy' as developed in the arguments of Appadurai (1992) and Lash and Urry (1994). Their approach provides a radicalization of the globalization debate beyond Gidden's understanding of it as a 'consequence of modernity'. Instead of stressing the continuity of the

9

institutional 'carriers' of modernity into late modernity, they emphasize a new velocity and flexibility of the internal structures of global society and its rather disorganized character. Their portrait of a global society is one of a landscape of non-isomorphic and intersecting global flows of people, commodities, finance, new technologies, knowledge, narratives, and symbols. The general flexibility and speed of these flows, as well as their disjunctive order, generates new forms of social and economic relationships, political alliances, and cultural identities, distinguished mainly by their 'deterritorialization' and the conflicting tendencies of homogenization and heterogenization (Appadurai 1992: 301ff.). In this sense, then, global society has been opened up again, both in its trajectory and outcome, instead of just drawing modern society to a close.

This radicalization of the debate on globalization is reflected in the transformation of the debate on world cities towards a new concept of the global city. Initial attempts to understand cities within a global perspective emphasized world cities 'as command and control centres in the capitalist world economy'. Instead of one metropolis providing the 'urban centre of gravity' of respective world economies, there was now a single world system controlled by a global network of world cities (cf. King 1991: 12ff., 24ff.). According to King, it was still a 'major paradigm shift' when during the 1980s European urban studies, induced by the evidence of global interdependency in the form of European metropolitan regions losing jobs to cheap labour regions across the world, recognized that urbanization processes both at the periphery and the core of the world economy are shaped by the *same* global economic forces beyond national control (1991: 3ff.). What was happening during the 1980s was a radical restructuring of the world economy, both in technical and geographical terms. This restructuring can be largely described as decentralization and deterritorialization, made possible by new technologies, and resulting in a 'general globalization' of capital and economic activity (Sassen 1991: 22ff.; King 1991: 90ff.). To put what is a fairly complex transition into simple terms, an international world economy characterized by its territorial organization, was replaced by a global economy determined by more flexible 'flows' of finance, commodities, technology, and information.

In this new configuration of global flows, major cities acquire a new strategic function. Increasingly 'unhooked' from their respective national hinterland, they serve as the relatively freestanding 'nodal points' of those global flows, or more precisely, their coordination, control, and partly also their production. According to Sassen (1991: 3ff.), this new type of city is characterized by the paradox of concentrating the control over global resources on the one hand, while on the other hand global capital and finance have radically restructured the economic and social order of these cities. In other words, the global city is a centre of global activities. However, unlike the world city, which was determining the fortunes of modern (world) empires, the global city's 'fortunes [are] decided by forces over which it has

10

little control' (King 1991: 145ff.). A point that is further elaborated in Budd and Whimster's collection of essays concerned with the impact of global finance on a multiplicity of aspects of London's socio-economic make-up (1992).

To bring home this line of argument, which stresses difference rather than continuity, just as the territorialized international world economy is not the same as a global cultural economy of intersecting flows, so is the world city – functioning as the hegemonic gravitational centre of regional world economies or empires – not the same as the global city, characterized as a key node of global flows, which it concentrates and bundles, but not entirely controls. It goes without saying that these two concepts are ideal types, just as the transformation of a world city into a global city does not happen overnight. This, however, does not mean that one cannot identify mile stones in this transformation. As for London, the transformation and globalization of the financial markets in the 'Big Bang' of 1986 certainly had such significance (cf. King 1991: 90ff.; Jessop and Stones 1992: 171ff.).

It is not my intention to take sides between these two lines of argument, but merely to show that there is good reason to assume an important role for the urban as a carrier of globalization processes. However, it is equally plausible to assume that the urban in the configuration of a global society will undergo transformation. Indeed, if the number of people living in urban places will have doubled by 2025, then it makes sense to talk with Clark (1996: 166f., 187) of a 'global urban society'. It is equally important to foresee that the urban world of global society will not look the same as the one that developed during urbanization in modern society. Like other aspects of modernity, with the onset of the 'Global Age' (Albrow 1996), both the urban and the process of urbanization itself are moulded into a new configuration of social order, interaction, and attitude, and thereby more or less transformed. The reshaping of the metropolis into a global city is certainly an indicator of that change. Moreover, the most pressing issue in which the new configuration of globality already surfaces in the urban world is the concern about the sustainable future of the global city (Harvey 1996: 426ff., 434ff.; Cadman and Payne 1990).

From this argument two implications need re-emphasizing in order to signpost the further approach of this study. First, major cities within global society might have lost their undisputed status as urban centres of gravity that exercise hegemonic power over their surroundings, and yet, it is exactly for that reason that researching the global city's changed status and transformed internal make-up can provide indicators for the transition of society at large. Second, if the global city is to be seen as a 'prism' of change and transition, then it is certainly not just its socio-economic make-up and its ecological sustainability that need investigating. First and foremost, the new configurations of an emerging global society and culture should be equally detectable in the everyday life of a city like London. Relatedly, the lives and narratives of ordinary Londoners should tell us as much about London as a global

11

city, as would the number of headquarters of top financial institutions present in the City of London.

The global city and its microglobalized hinterland

What is striking when looking at the plethora of literature on the global city, is the fact that very little has been researched or written regarding the life world of the global city. Perhaps this is not surprising, considering that the debate on global cities/world cities, like the wider discourse on globalization, emerged from world system theory. Accordingly, emphasis in these studies is given to the linkage between global cities such as London, Tokyo, New York and Frankfurt as centres of global banking and finance, and subsequently, to the manifestations of this function in their respective socio-economic, institutional, and even cultural environment (e.g. Sassen 1991, 1994; King 1991, 1996; Budd and Whimster 1992; Knight and Gappert 1989). None of these studies tackle the manifestation of globalization processes in the everyday lives of 'ordinary' global city dwellers. As far as *Living the Global City* is concerned, Eade's (1997) collection of that name remains a welcome exception; otherwise, reading Hanif Kureishi's novel *The Buddha of Suburbia* (1990) is still as enlightening as the academic literature. It is this largely empty space in the debate on global cities, regarding the life-worlds of its inhabitants, that this book hopes to help fill.

What is striking on second sight, is the terminological confusion prevalent in the debate on global cities. The range of terms applied to the global city stretches from 'major cities', 'premier cities' in Sassen's analysis (1991), 'world cities' and 'supranational key cities' in King's account on London (1991), back to 'metropolis' in Budd and Whimster (1992) or 'transactional metropolis' in Jones (1990), and on towards the more cautious and ambiguous attempts of 'so called world cities' in Harvey (1993) and the rather diplomatic notion of 'cities in a global society' in Knight and Gappert (1989). Most importantly, there seems to be no clear differentiation between world city and global city, so that most accounts on global cities feel free to 'use the terms "world city" and "global city" interchangeably' (Brenner 1998: 29 endnote).

This relaxed handling of terminology is certainly due at the least to a sloppiness of thinking. Obviously, different approaches towards a dazzling phenomenon such as 'cities in global society', will bring different aspects of it to the fore, and accordingly require distinctive terminologies, in parts complementary, partly overlapping, or even contradictory. Accordingly, it would hardly be worth raising this terminological problem in the context of this book if it was not exactly this interchangeable use of the terms 'world city' and 'global city' on the one hand, or the often undifferentiated use of the term global city, which obscures the life-world dimension of the global city phenomenon. It is on those grounds that for the purpose

12

of clarifiying the argument of this book, the following terminological distinctions are suggested.

If we assume major cities are at the centre of the 'global cultural economy', as described by Appadurai and Lash and Urry, then they are certainly not just nodal points of flows of capital, finance and information, but also central to the flows of people and their social practices and beliefs. It is in the latter sense that these cities need attention as 'great cross roads' of different life-worlds and centres of 'world-mindedness' (Gottmann 1989: 62ff.). To re-emphasize these different dimensions of contemporary major cities it appears useful to clarify initially the analytical distinction between 'global city' and 'world city'. The term 'global *city*', with the emphasis on the city as the centre of financial and other related transactions, shall be reserved for the socio-economic function of major cities. While the notion of '*world* city', with the emphasis on, literally, the presence of the world in one city, and the world-mindedness of those cities, shall be used when referring to the make-up of the life-world of major cities.

Of course, these are ideal distinctions, and not definitions of absolute validity. To illustrate these analytical distinctions with examples: London is certainly both a 'global city' and a 'globalized world city', while Frankfurt is a challenger for 'global city' status, but can hardly claim to be a 'globalized world city' in comparison with places like London or New York. While London might lose out to Frankfurt or Paris regarding its 'global city' function in Europe (Budd and Whimster 1992: 18), it will certainly remain a global 'cross road of life-worlds' for the foreseeable future.

Still, this distinction needs specification, as soon as historic developments enter the picture. While the 'global *city*', in the way we just defined it, is a fairly recent development, the '*world* city' in the above sense is not. It is the outcome of a long-standing process of integrating the world's cultural and social variety, with all its challenges and contradictions, into one city. Thus, living in the world of everyday life of such a city means potentially living *the* world in one city. It is in this sense that Goethe supposedly applied the term *Weltstadt* as early as 1787 to Rome and later on to Paris (Gottman 1989: 62). And, as Dr S. Johnson, the famous eighteenth-century London writer, has already remarked:

> Sir, you find no man, at all intellectual, who is willing to leave London. No, Sir, when a man is tired of London, he is tired of life; for there is in London all that life can afford.
>
> (cit. in Tames 1992: 102)

While this process of bringing together cultural goods and social practices from different corners of the world continues, the configuration in which this process takes place has changed. It is argued that not only have globalization processes transformed major cities like London in their socio-economic role and function, turning some of them into 'global cities', but have had an equally radical impact on the life-world dimension of those cities. It is in this sense that the 'world city' becomes a *'globalized world city'*; its life-world displays the new patterns of deterritorialized global social and cultural flows.

The processes that Harvey aptly summarizes under the term 'time–space compression' have radically changed the geography of places, people, symbols and practices. This has resulted in their radical mobilization and the subsequent replication of social worlds across the globe, making it rather difficult to allocate them to an initial context of belonging. The accelerated 'implosion' of the world's socio-geographic spaces, following Harvey's argument, means that different social and cultural worlds 'collapse upon each other' in significant places such as the 'so-called world cities'. The internal make-up of the globalized world city is then best described, in a first approximation, as a 'collage world', a world of juxtaposition rather then integration (Harvey 1993: 240ff., 294ff., 299ff.). In line with Harvey's argument it can be maintained that, metaphorically speaking, just as the 'global city' is largely the outcome of 'deregulation' within global capitalism, so is the 'globalized world city' a result of the 'deregulation' of the spatio-cultural patterns of the life-word. Deregulation of the life-world emerges, for instance, in the peopling of London (Merriman 1993). London's immigration patterns continue to reflect the long-established links between the former 'world city' and what used to be the British Empire (King: 1991: 139ff.). As a 'globalized world city', however, contemporary London's immigration patterns increasingly reflect the global 'ethnoscapes' (Appadurai 1992) of a largely de-territorialized and disjunctive global cultural economy. London's comparatively large Latin American community, which only really came into existence from the 1970s onwards (Merriman 1993: 149ff.), is an indicator of these new links that extend beyond the traditional connections between London and people from the Commonwealth countries. Deregulation, in the above sense, also took place at local level, insofar as most London localities can no longer be fixed in their ethnic and social character, but instead accommodate a juxtaposition of different cultural and social worlds (Budd and Whimster 1992: 19).

Such coexistence of different social and cultural worlds in one locality – with their respectively spatio-temporal practices overlapping and intersecting, but hardly coinciding, and in any case extending beyond the actual locality – has been aptly described by Albrow (1997) as a landscape of 'sociospheres', in short, a 'socioscape'. In keeping with this argument, the life-world of the 'globalized world city' is then possibly best described as a configuration of 'sociospheres', each of which is to a higher or lesser degree linked to the wider landscape of global cultural flows.

Accordingly, and slightly opposed to Harvey's postmodern emphasis on collapsing and imploding spatialities, the 'globalized world city' sees the compression of global cultural and social flows into its configuration of coexisting sociospheres, thereby modifying but not dissolving them. Such a compression and, subsequently, accommodation of global variety and difference into a distinctive sociocultural environment will be defined as 'microglobalization'.[2] In this definition, the prefix 'micro-' does not imply 'macro-globalization' as its counterpart, but refers to the intense globalization of the world city's everyday life, transformimg it into a 'globalized world city'.

Three different meanings are integrated in the complex notion of the 'microglobalization' of the world city. First, it makes the 'globalized world city' the embodiment of the globe's social and cultural variety. As such the 'globalized world city' retains the role of a 'metropolis', in Mumford's sense of providing 'the biggest aggregation of human life, the most complete compendium of the world' (Mumford 1991: 639). However, this notion applies with the qualification that it accommodates rather than absorbs and dissolves these differences. Second, the world's variety and complexity is not just simply re-presented in a microglobalized environment, but is compressed in the relative spatial proximity of a bounded local social setting. This intensification of metropolitan life encourages the emergence of new social forms, and allows us to study them in the making. Hence Park's metaphor of the city as a *laboratory* of social developments (cf. Lindner 1990: 90) is especially appropriate for the 'globalized world city'. Third, the 'globalized world city' is a symbolic '*microcosm*'. The variety and complexity of the microglobalized environment of the world city is an everyday experience for the people who live in it, and who consequently have, to some extent, to make sense of it as their world of everyday life. Their daily routines draw upon, generate and (re)shape the reality of the globalized world city. The attempt by different individuals to mark out the internal structure of their 'soft city' through different kinds of symbolization and cate-gorization illuminates the 'globalized world city' with meaning from within, thus creating what Schütz calls a 'cosmion' (1967: 336). Moreover, as the everyday lives of people in the 'globalized world city' are linked in many ways to people and places all over the world, the 'cosmion' of the world city reflects these external global links in its internal symbolic structure, making it a *symbolic microcosm of the global cultural economy*. It is this 'cosmion' of the 'globalized world city', which this book will further explore through the analysis of the narratives of eight Londoners.

2

TOWARDS A PHENOMENOLOGY OF GLOBALIZATION

No spot on this globe is more distant from the place where we live than sixty airplane hours.

A. Schütz (1971: 129)

A phenomenology of everyday life

It is not my intention in this chapter to marvel, with the benefit of hindsight, at the interesting things Husserl and Schütz would have to say about globalization, though it is interesting to note how many of the issues of the contemporary debate on globalization are there implicitly, particularly in Schütz. However, it is the intention of this book to take a step forward towards what could be termed a 'phenomenology of globalization'. This term needs explaining. In the way it is used here it has nothing to do with phenomenology in strict Husserlian style. That is, nothing to do with an attempt to find the transcendental foundations of global consciousness, or to discover the 'essence' of globalization by bracketing the complexity of its many manifestations in everyday life through phenomenological *epoché*. Instead, in a much broader sense, it refers to such a 'phenomenological description' that provides 'a severely empirical register of what is happening to the frames of meaning in people's lives as globality, globalism and globalization take hold' (Albrow 1996: 79). Rather than retreating from the practices of everyday life into abstract philosophical exercise, this type of phenomenology engages with everyday life in order to grasp indicators for new forms of people's engagement with a rapidly changing world. The only methodological pre-assumption in this approach would be that if there is something like a 'global age' approaching, then 'we can listen to the new age on the street' (ibid.: 80). The phenomenology of globalization proposed in this book is an empirically grounded phenomenology of everyday life,[1] and in this sense perhaps closer to Socrates than Husserl.

Obviously, the ways in which globalization surfaces in people's accounts of their everyday lives are as manifold and divergent as the ways in which globality and globalization take hold of contemporary life. However, carefully listening to the narratives under closer investigation in this book, one theme that runs through all of them could be named in a first rough approximation as disembedding and re-embedding of *Umwelten*. What is being articulated is that people's familiar routines, their feeling of belonging, and their meaningful relationships to significant others, are increasingly delinked from the primacy of locality and local community. Relatedly, there is a complementary story-line articulating the individual's active effort to generate and maintain situatedness and familiarity within a field of action and experience that stretches across distance and beyond the definition power of local settings. Without much theorizing, it thus seems obvious that the restructuring of people's everyday lives across distance brings to the fore much more prominently the individual phenomenologies that are trying to make sense of these processes. This makes these individual phenomenologies the starting point for any further theoretical investigation into processes of disembedding and re-embedding of everyday life in the context of globalization processes in the global city.

To be precise, it makes them the *logical* starting point for an empirically grounded phenomenology of globalization. Even though this approach emphasizes the primacy of narrative accounts about living in the global city, it would be naive and misleading to assume that the way it reads and interprets those individuals' narratives would not at the same time reflect theoretical accounts about globalization. Regarding the particular theme under investigation here, it reflects two core ideas of the globalization debate. On the one hand, the theme of the disembedding and re-embedding of people's *Umwelten* clearly implies deterritorialization, which is, according to Appadurai (1992) and Lash and Urry (1994), *the* central force behind the emergence of a global cultural economy. This force also takes hold of people's most familiar settings and meaningful practices and thus challenges the situatedness of their everyday lives. Insofar as it pursues the different ways in which familiarity, competence and normalcy are being maintained in people's everyday lives across various time–space distances, the phenomenology of globalization proposed in this book could perhaps be adequately labelled as the 'phenomenology of lived spaces or a phenomenology of time–space distanciation'.

The narratives under investigation here stress the individual's effort to re-appropriate meaning and familiarity in a world that is seemingly escaping their control, and highlight the biography as the primary frame of reference for generating and structuring lived space. There is evidence in these narratives for unrelated individual worlds, coexisting next to each other within the global city, more or less disentangled from local social and symbolic discourses, while at the same time increasingly participating in global cultural flows. This underlines processes of individualization as complementary to the process of globalization, in so far as the latter continues processes of de-traditionalization by setting individuals free from traditional frames of reference, such as local community and even family. This idea has gained prominence mainly through Beck, who has argued that 'world society becomes a part of biography' and the individual in turn becomes the 'actual reproduction unit of the social in the life world' (1992: 130, 137). It is also evident in Appadurai, who sees 'the individual actor as the last locus of this perspectival landscape' of global cultural flows (1992: 296), and Melucci, who maintains that in times of planetary complexities the individual's biography 'still provides one of the principal reference systems' for the construction of a person's identity (1996: 105, 43). Insofar as the following investigation pursues the issues of lived space and situatedness mainly from within a biographical context, and clearly stresses the individual's active effort to find a sense of place and belonging in a transitory and expanding global environment, it could be described as a phenomenology of individualization.

A phenomenology of extended milieux

It is one thing to follow people's narratives on disembedding and individualization in an ethnographic description. It is quite another task to explore systematically the conceptual implications of the themes running through these narratives. In other words, just as people talk in new ways about their lives in a world that is caught up in a far-reaching transition, the sociologist needs ways of conceptualizing and writing that reflect these changes (Albrow et al. 1994). It is only at this point that a phenomenology of globalization would draw on phenomenological theory in the narrower sense. For, with its overall emphasis on life-world rather than territorial units, and with its priorities given to the individual's frames of meaning and action rather than the locale, it would seem that phenomenology proper has something to offer in respect of adequately conceptualizing processes of disembedding and individualization as they surface in people's everyday lives. The phenomenological concept that is particularly suited to grasp these two complementary processes is that of *milieu*. While initially developed by Max Scheler in the context of German phenomenological anthropology, the idea is also present in Schütz's concept of the 'biographical situation', Goffman's notion of *Umwelt*, and Cassirer's theory of 'symbolic space'. As the reader will see, the way I want to use the concept of milieu in the following analysis makes use of the complementary analytical potential of each of those ideas for the investigation at hand, and is less concerned with the consistency of their phenomenlogical assumptions or their coherence within phenomenological theory.[2]

Accordingly, an initial working definition is inspired by each of the aforementioned ideas. *Milieu* shall be defined as a relatively stable configuration of action and meaning in which the individual actively maintains a distinctive degree of familiarity, competence and normalcy, based on the continuity and consistency of personal disposition, habitualities and routines, and experienced as a feeling of situatedness (cf. Grathoff 1989: 344, 434).

From Scheler this definition takes the idea that the milieu is the individual's value-related disposition towards the world, as such providing a 'filter' or an 'alphabet' through which the world is perceived, but at the same time ordered and arranged for practical intervention. Important in Scheler's account is the differentiation between 'momentary milieu' and 'milieu structure'. While the individual's milieu structure remains relatively stable despite changing environments, the momentary milieu is the current and transitory content of the actual environment, which is practically relevant at any one moment, filtered through the individual's 'order of values' (Scheler 1973: 130ff.). What follows from Scheler's ideas is that we carry our milieu around with us, attempting to realize it wherever we came to live. Relatedly, people might share the same physical environment, and yet practically live in different milieux. Finally, people with the same milieu structure might relate to each other on the basis of like-mindedness rather than a shared locale.

From Schütz we adopt the idea of 'relevancies', that is the ordering of the individual's environment into spatio-temporal segments that are relevant practically to the individual's varying tasks at hand. The system of relevances that frames the individual's everyday life carries the index of his or her 'biographical situation' – life plans, projects, skills and abilities, and corresponding stocks of knowledge. Of particular interest here is Schütz's assumption that the zones of practical interest and concern are by no means confined to the immediate surroundings. Through technical devices, the individual's 'manipulatory sphere' is constantly expanding, linking the individual's 'here and now' to spatially distant events and people. Accordingly, the individual's relevant environment is not organized in concentric circles of decreasing intimacy and familiarity, but should rather be imagined as a patchwork of overlapping zones with non-rigid boundaries and open horizons, kept together by the individual's determination to pursue long-term life plans and tasks at hand. In this sense Schütz can argue that the 'biographical situation' provides the basic frame of reference and orientation that guides the individual's attempt to generate familiarity and competence in a complex and expanding environment (cf. Schütz 1970: 167ff., 1966: 121ff.; 1967: 306ff.). What Schütz provides us with is a model for the concrete structuration of the life-world, which puts the individual at its centre, and in which proximity and distance are measured according to biographical relevance rather than geographical distance.

Goffman contributes a further clarification regarding the link between the individual's immediate surroundings and distant zones of concern and interest. In his concept of *Umwelt*, Goffman differentiates between 'personal space', that is the individual's immediate surroundings, and the individual's 'fixed territory', referring to fixed belongings and possessions in distant regions of the life-world. In so far as the individual has arranged this fixed territory in a unique way, it is an essential symbolization of the 'self', and therefore effectively part of the individual's *Umwelt*, regardless of spatial distance between the self and its fixed territory. Conversely, elements of the fixed territory, such as the telephone, can play the important role of 'access points' to the individual's *Umwelt*, indicating sources of interest or alarm long before these reach the actual personal space. What links the two components of *Umwelt* in Goffman's approach, is that they are actively designed to preserve the individual's sense of competence, normalcy and control through maintaining personal standards of order and conduct. Moreover, this concern with the maintenance of personal standards of conduct and normalcy brings another of Goffman's ideas into consideration. Increasing individual mobility, amongst other things, implies detachedness from home and other components of the fixed territory of the self, in turn requiring the ability to generate home-like conditions wherever necessary. Goffman's notion of 'back region' seems to grasp this problem when interpreted as a region in public that the individual claims and transforms for purposes of relaxation and recuperation from the demands of public norms (Goffman 1972: 52ff., 337ff., 351ff.; 1990: 109ff.).

Finally, from Cassirer we take on board the important insight that lived space is always 'symbolic space' in at least two ways. First, lived space is not defined by geometrical determination, but is a dynamic unity, defined by the overlapping and contradicting directions of action in any one situation. Moreover, these actions find their direction not just by rational calculation, but are inspired by hopes, fears, and anxieties. In this sense, lived space is not just a dynamic field of action, but filled with the atmosphere of contradicting human affects.

Second, lived space is perspectival space, depending on the point of view in space and also time. This implies that lived space is not confined to the actual here and now, but represents 'the present and the non-present, the real and the possible'. The individual, able to distance him or herself from the immediate surroundings, grasps his or her own position as being relational to a larger spatial 'order of possible coexistence', and is able to anticipate future possibilities as to how the environment could unfold in relation to one's own life plans (Cassirer 1953: 212ff.; 1957: 152ff.; 1970: 25ff.). Thus, the ability to grasp lived space as symbolic space is an essential precondition for successfully completing and, when necessary, modifying a project or task amidst changing and increasingly complex spatial circumstances.

What these different contributions to the milieu concept have in common is their emphasis on non-localized forms of situatedness, generated by the individual's active effort to maintain competence and familiarity in a world not of its own making. In this version of the milieu concept, situatedness is not bound to a particular locality, it has no strict boundaries, and its territoriality is a function of the individual's practical relevances. This clearly sets it apart from concepts that emphasize the primacy of place in the organization of situatedness, and which stress the shared culture that derives from inhabiting the same locality or territory. It is in this regard, I would maintain, that the concept of milieu is better equipped to conceptualize processes of disembedding and individualization, as opposed to, say, the concept of community. Accordingly, it can be argued that a phenomenology of globalization, as proposed in this book, would, at its core, have to be a 'phenomenology of the milieu'.

This is not to say that the very processes of globalization, which it helps to conceptualize, would not at the same time pose a challenge to the concept, demanding necessary refinements of its assumptions and implications (Albrow et al. 1994; Keim, 1997). First, living in a global society, especially the global city with its plethora of life styles, poses a challenge to the individual's order of values and relevances. In a life-world of huge social and cultural complexity and contradictions, the search for situatedness is likely to become, more than ever before, a continuous process, rather than a final achievement. 'Situatedness' would then, more than anything else, refer to the individual's ability to handle the unexpected, rather than the effort to guarantee the expected. People's geographical mobility and the access to global means of communication pose questions regarding the structural make-up of the milieu. Perceptual field, effective field, field of action, and manipulatory sphere will no longer necessarily centre around the body-related 'here and now'. We might be physically located in a certain setting, but our emotions are tied to somewhere else, and our actions are geared towards yet another region of the globe. This tendency might be decribed as a 'defocusing' of the milieu, insofar as the effectively experienced 'here and now' extends beyond the perceptual field. More provocatively, it could be suggested that there is no solidly defined 'here and now' any more. Relatedly, the notion of situatedness might become more closely linked to distant fragments of the fixed territory of the self, rather than being associated with the individual's immediate *Umwelt*. Furthermore, as global media of communication such as telephone, fax and e-mail become familiar parts of the milieu, rather than simply 'access points' of an outside world, they encourage us to rethink the milieu concept's implications of 'familiarity' (with distant 'relevant' localities), and 'normality' (with 'relevant' contemporaries) without unduly relying on presence and co-presence respectively. Finally, individual mobility, no longer confined to elitist transnational cultures, might imply tendencies towards the 'generalization' of milieux. Travelling between significant places and relevant others, people will have to rely increasingly on formalized settings, such as fast-food outlets, airports, car rentals, petrol stations, and so on, which serve the basic needs of different people in a standardized way almost anywhere on the planet.

The implications outlined could then be summarized in the notion of *an extended milieu*. Milieux become extended in a plain geographical sense through individual mobility and the use of global means of communication. They extend in the sense of incorporating and familiarizing new technologies and standardized facilities as means of connecting distant fragments of the territory of the self. Milieux also extend insofar as care and maintenance work within the milieu cannot be limited to the individual's immediate presence, but will increasingly rely on forms of familiarity and control mediated through distant 'relevant others'. Finally, in a more philo-sophical sense, these developments are likely to foster a broadening of people's horizon. As it is these ideas and questions that are at the core of the argument pursued in this book, the approach it follows is best described as a 'phenomenology of extended milieux'.

Re-locating the concept

While it would seem that at present such a phenomenological approach is largely missing from the debate on globalization as well as global cities, it has in parts been explicitly claimed or implicitly assumed elsewhere.

The idea of a phenomenology of globalization has its forerunner in Berger et al.'s *The Homeless Mind*, a study that claims to ground its phenomenology of modernization in the 'reality definitions' of everyday life consciousness (1974: 18). However, whilst I am in agreement with their general phenomenological approach, the analysis in this particular study comes to rather different conclusions. This should not come as a surprise as, after all, we look at two different epochs between which there are continuities (see Giddens 1994), but certainly also ruptures (see Albrow 1996). While Berger et al. detected the 'discontents' of 'homeless minds' from their analysis of the modern life-world (1974: 163ff.), the narratives under investigation in this book allow for the possibility of extended milieux, not caught between uprooting and subsequent attempts of re-rooting, but fairly capable and reasonably happy to live across distances and between places.

More recently, it was Giddens who explicitly proclaimed a phenomenological perspective within the actual debate on globalization. Though he calls it a 'phe-nomenology of modernity' (1994: 137ff.), it is *de facto* a phenomenology of globalization, in as much as the globalization of people's 'phenomenal worlds' is a crucial aspect of the transformation towards what Giddens defines as 'late modernity' (1993: 181ff.; cf. Tomlinson 1994). On inspection it becomes clear that there are close parallels between the field of research sketched out for a phenomeology of extended milieux in the previous section, and the issues raised by Giddens for a phenomenology of late modernity. The most obvious ones include: 'dis-placement', 'phantasmagoric' places, familiarity 'mediated by time–space distanciation', 'intrusion of distance into local activities', 'centrality of mediated experience', subsequently radical extension of the individual's experienced phenomenal world beyond the immediate *Umwelt*, to mention the most obvious ones (Giddens 1993: 187f.; 1994: 137ff.). There is less correspondence as to what the actual phenomenological approach that attempts to answer these questions should be like. Giddens' phenomenology is hardly empirically 'grounded' in people's actual life-worlds. Instead, he absorbs these important analytical questions concerning the actual restructuring of people's milieux into the rather metaphysical metaphor of everyday life in late modernity being a 'juggernaut' over which the 'ordinary individual' loses control. The 'phenomenology of late modernity' is reduced to the question 'What does it feel like to live in the world of modernity?', and is thus aptly named as 'existential phenomenology' (1993: 35ff., 50f.; 1994: 137). As such, it could be seen as being closer to Berger et al.'s phenomenology of modern consciousness, than to a phenomenology of extended milieux. While Giddens' emphasis is on an 'overall picture of the psychological make-up of the individual' living in late modernity (1993: 35), the argument of this book is predominantly concerned with the spatio-temporal restructuring of the individual's milieu.

Robertson, on the other hand, as the other main figure in the debate on globalization, rejects the phenomenological approach outright. He regards the phenomenological approach as a main contributor to the 'nostalgic paradigm', which he exemplifies with Berger et al.'s 'homeless mind'. Robertson's criticism of the phenomenological approach is directed towards the image of 'home' being a place of 'refuge' from an alienated world (1992: 156). The phenomenology of extended milieux pursued in this book rejects the nostalgic parochialism that some phenomenological approaches are rightly accused of. Instead, it stresses the individual's active engagement in the generation and transformation of a global life-world. The detailed investigation of extended milieux thus contributes to the theoretical analysis of the 'concrete structuration of the world as a whole', beyond systemic forms of integration, and 'relevant to the world in which we live', as asserted by Robertson. Moreover, the extended milieu can be seen as one possible way in which the individual positions him or herself 'in relation to the global human circumstance'. In turn, the extended milieu then becomes a crucial building block of the 'global field' itself (Robertson 1992: 25ff., 51ff., 61f.). From this perspective it would seem that a phenomenology of extended milieux would offer a crucial complementary element to Robertson's comprehensive theory on how global society and culture are ordered in a meaningful way.

As far as the theoretical core argument of the phenomenology of extended milieux is concerned – namely the delinking of milieux in general, and their situatedness in particular, from locales and localized social relationships – there are parallels once again, though more explicitly expressed in Giddens' than Robertson's theories. Consistent with his concern about the parochialism attached to the idea of locality and local culture, Robertson argues that 'in the present situation of global complexity, the idea of home has to be divorced analytically from the idea of locality' (1995: 39). In Giddens' theory of globalization, the core idea of 'disembedding', though initially focused on the time–space distanciation of the institutional clusters of modern nation state society, clearly extends towards everyday life. The idea that disembedding mechanisms lead to people and their social relations being 'lifted out . . . of their "situatedness" in specific locales', and subsequently extended across time–space, is central to Giddens' line of argument at various points (1994: 52f., 18ff., 108f.). However, in Giddens' understanding, the extension of people's environments is largely due to the disembedding logic of modern institutions, exemplified in the 'abstract systems' of money and institutionalized expertise. It is then only consequent for Giddens to assume that the extension of people's milieux must feel like 'being aboard a careering juggernaut' on which they have little or no control (1993: 16ff.; 1994: 53). Hence the 'existential phenomenology' attached to Giddens' theory of globalization.

With regard to the conceptual consequences of the disembedding and extension of people's milieux, a similar ambiguity becomes obvious. Rather similar to the earlier proposed modifications of the milieu concept towards the notion of an extended milieu, Giddens suggests a linking of Goffman's idea of *Umwelt* and Schütz's concept of 'system of relevances' (1993: 127f.). However, Giddens' analysis remains ambiguous in this regard as he uses 'milieu' quite frequently, though in a rather common sense way, but when actually taking a phenomenological stand, he tends to revert to Goffman's understanding of *Umwelt* as a 'protective cocoon' (1993: 83f., 97, 112, 126ff., 182). It is one of the aims of this book to make a contribution towards a clearer usage of these phenomenological concepts.

3

EIGHT LONDON BIOGRAPHIES – LIVES IN THE GLOBALIZED WORLD CITY

> I have often amused myself with thinking how different a place
> London is to different People.
>
> James Boswell (cit. in Shakespeare 1986: 10)

London has always meant different things to different people, depending not only on their reasons for coming into the metropolis, but also on the whole configuration of circumstances under which they then had to make the big city work for them. Different incomers and newcomers, both in the past and more recently, would tell very different stories about living the big city (Merriman 1993; Shakespeare 1986). The eight biographical sketches introduced in this chapter are no different in this regard. In its own way each of them reflects life in contemporary London.

However, there is more to these sketches than just a further addition to the variety of recorded London life stories. These biographical portraits serve an argumentative purpose, and have been selected with methodological consideration. Based on semi-structured biographical interviews, they reconstruct the 'biographical situation[s]' (Schütz) of eight Londoners, not in their entirety, but insofar as they reflect the de-linking of milieu and locale, and the subsequent 'extension' of those milieux both within and beyond the life-world of the global city. In other words, these sketches provide the biographical reference points in which most of the analytical 'themes', which will unfold across the following chapters, are 'grounded'. The sketches also serve to provide the reader with the opportunity to relate narrative sequences used within the analysis of the following chapters to the biographical context from which they have originated. The attentive reader will then realize that the narrative accounts in the text stand not in isolation, but serve to 'ground' the different dimensions of the concept of an 'extended milieu'.

Accordingly, the eight portraits of London-based milieux have not come about by randomly interviewing a group of punters. Although they deliberately represent different biographical and social worlds, they have been carefully selected in relation to each other, following principles of grounded theory, such as 'maximizing differences' (Glaser and Strauss 1979; Strauss and Corbin 1990). Thus, while, on the one hand, each of these portraits touches on themes such as access to London, a sense of belonging, social contacts at home and at work, engagement with London's social and cultural variety, mobility in and beyond London, links to distant places and people, and so on, on the other hand, these portraits highlight the, sometimes extremely, different ways in which these issues are perceived and handled in each of the eight milieux under investigation. So, while biographical in content, each of these portraits also provides a clear case of some of the different types of milieux that can be found in the globalized world city. To indicate this, each portrait carries both a personal name and an ideal typical characterization. Finally, taken together, these portraits should provide the reader with a first ethnographic approximation to the overall theme of this book: extended milieux interwoven with the microglobalization of London's everyday life.

Such a biographical approach is not without its critics. For the investigation at hand, however, it is the most appropriate one. While phenomenological sociology from Scheler to Schütz has always insisted on the individual as a 'distinctive dimension of meaning in the life-world' (Grathoff 1989: 135f.), this has to apply in particular to a phenomenology of globalization as proposed in the previous section. First, because it attempts to ground its analytical argument in the empirical investigation of what happens to people's milieux in the context of globalization and globality, something best accessible through people's narratives about their biographical situation within that global configuration. Second, this biographical approach has, if anything, been re-emphasized by the recent debate on individualization, regarded as *the* social process complementary to globalization. This debate has proclaimed the 'individual's return to society' (Beck 1997: 94; see also Beck 1992; Melucci 1996). At its core the individualization thesis is based on the assumption that people are being set free from traditional frames of action and meaning, so that subsequently 'the individual himself or herself becomes the reproduction unit of the social in the lifeworld' (Beck 1992: 130).

Accordingly, the concept of milieu pursued in this book is explicitly focused on the individual's attempt to generate and maintain familiar surroundings. This is consistent with the emphasis given to the individual actor in each of the theories that inspired this particular notion of the milieu: Scheler's initial concept of the milieu, as distinct from the collective horizon of the life-world; Schütz's notion of a biographical situation within the social world shared with consociates and contemporaries respectively; and finally Goffman's *Umwelt* as territory of the self. This brings the milieu concept brought forward in this book rather closer to Beck's and Melucci's individualization thesis, and at the same time differentiates it from Bourdieu's notion of a collectively reproduced habitus. However, to re-emphasize a point made in the introduction: the assertion of a coexistence of relatively unrelated

26

milieux reflects *a* lived reality as can be experienced in the globalized world city, but does not attempt to make claims regarding *the* lived experience of global city life.

Sarah Penhaligan – the 'mobile cosmopolitan'

Sarah, in her thirties, has been living in London on and off for six years. Brought up in a middle-class family in Surrey, just outside Greater London, she used to go up to London's West End, either with parents or later with friends. After finishing her studies at Oxford she lived in London for three years, working as a stewardess for British Airways. While still working for British Airways she bought a house in Oxfordshire and commuted to London. She then started working as a head-hunter for a personnel recruitment agency based in the City of London, and decided to buy a house in Peckham, South London, where she has been living for three years now. Her job with British Airways took her to places around the world, and Sarah stresses that she saw this job as an opportunity to see the world rather than earning money. Whenever possible she would volunteer for flights with time off at the destination before returning to Heathrow. But, she insists, unlike other crew members, she tried to be 'incredibly adventurous' in order to not let herself be confined to the airport and hotel setting. In her current job she recruits personnel for international banks and security houses, 'moving people between Paris and Frankfurt, between Frankfurt and London'.

While her work means that she is firmly located in London, the money she now earns in the City, and also the social contacts she makes in that environment, allow her to live a fairly mobile and cosmopolitan life style. Sarah casually mentions 'weekends in places like Boston or Barcelona, sailing weekends at the coast of France'. Also her hobby, singing in a church choir specializing in medieval music, has taken her to workshops in places like Venice. In this regard Sarah regards London as the place to be, because its 'communications are good'. Sarah also cites London's cosmopolitan character as one of the reasons 'why I like living here so much'. London, as she sees it, has so much to offer that 'there will always be someone or something you can join'. In comparison to, for example, Hamburg, where she lived for a substantial period of time, London appears to her to be 'a less conservative society', where one can get away with doing 'really eccentric things'. For her personally that means to dress in extremely different clothes when off-duty. Sarah also enjoys the fact that on a day out 'you can choose and eat anything from any part of the world', and Brixton for her is a place where one would go on a Saturday morning 'to buy exotic things from the market'. She describes her own neighbourhood as 'multicultural', referring to Nigerians, Ghanaians, Turks and Cypriots as her neighbours in the wider sense. What is more important to Sarah than their ethnic roots is the social background of these people. In this regard she believes they are 'people like me, . . . middle-class people with quite good jobs and very good education'. Sarah considers some of the Turkish and Cypriot people in her area to be 'far more Londoners than I am', as they have lived here for all of their lives. In general, she believes that 'there isn't such thing as a *real* Londoner'. Pressed in this direction she defines the 'Londoner' in a rather technical sense as someone who 'understands the system of living in London'. Although she feels that London is a very 'competitive' environment in business terms, and a place where you can live a 'fairly anonymous existence', it is also a 'mixed society' where people 'still tend to fit in and find something that suits them'. Personality-wise, she says it is very important to have the 'skills' to judge both people and situations rather quickly, and then to relate and socialize on that basis.

Her social life centres around work and her hobby, not the local community. Sarah's friends live all across London, and they socialize in a 'neutral territory', as she calls it, rather than entertaining one another at home. Consequently, 'home addresses aren't terribly relevant' for her social life, 'it's telephone numbers that are relevant'. Her friends are from different 'walks of life' and by no means would all know each other. This fragmentation of her social life is something Sarah actually encourages by 'dividing my friends into two sorts'. On the one hand 'friends who I work in the city with', whom she likes to go out with for entertainment, as they have 'a reasonable amount of money to spend'. On the other hand there are friends with whom she shares hobbies and interests, like former British Airways colleagues and musicians. Another way to stay in control of her complex (social) life is to never go out for a drink or meal straight after work, as lots of City people tend to do. Though this might be time-consuming, Sarah insists on going home after work in order to 'change gear' before returning to Central London for a meal or other entertainment. Sarah describes her neighbourhood as a 'friendly place', but says at the same time that there is no 'particular sense of community' any more, as most people, like her, are too busy working. Anyway, she chose this neighbourhood simply because of 'easy access to the City' (15 minutes by car), 'fresh air', 'lots of open space', and because property prices here were comparatively cheap. Sarah talks to one neighbour over the fence occasionally, while she does not know her other neighbour at all. 'Home' for Sarah is not necessarily linked to the local community in which she happens to live, but wherever her 'belongings' (books, records, pieces of furniture) are.

Sarah is clearly aware of her privileged social position that allows her a cosmopolitan life style in London. She says that there are others 'who might require more of a sense of belonging to one community, and not having outsiders'.

Ulla Anderson – the 'settled cosmopolitan'

Ulla, now in her eighties, is a 'Londoner by birth', from a family of mixed nationality. Her father's mother was Irish with French origins. She stresses that he was born in

Kennington, London to 'the sound of Bow Bells'. Being a paper merchant with business connections in Sweden, he met her Swedish mother on a boat returning from Sweden. The young couple finally settled in Streatham, South London, where her father built the house in which she now lives. While her older sister learned to speak both Swedish and English, Ulla was brought up with English as her native language. Still, she proudly refers to herself as a 'Londoner with a Swedish background'. Of three siblings, she was the only one to stay in London. Her older sister moved to Newfoundland, while her younger brother's career as a singer and actor took him to New York's Broadway. Due to the international extension of the family, and the general involvement in an artistic life style, mainly through her brother's friends, there was 'a rich social life' at her parents' home. Ulla believes that by then her family were 'more cosmopolitan than most people' in Streatham, where she had been asked on several occasions 'what do you speak in Sweden?'.

With her playwright husband she moved to the fashionable London suburb of Hampstead. Here she continued a cosmopolitan life style, entertaining politicians and actors not only from the UK. The couple also became good friends with their neighbours, Russian Jews from Austria. Ulla started a career as a model and dancer, but this ended with the start of the Second World War. During the war Ulla worked as a nurse. In this context she got to know a Polish Jewish woman, with whom she got on really well. This women now lives in Israel, but she and Ulla keep in touch via mail.

After her husband's death she returned to her parents' house in Streatham and started working as a teacher in South London. She recalls a significant change in the early 1970s, when all of a sudden 'you had something like seven different nationalities in the class, and you had to give them special English lessons'. She especially remembers pupils from Cyprus, and Asians from Uganda and Kenya, as well as Biafra. In this connection Ulla describes London as the 'the most truly free city', open for everyone, and open-minded. However, Ulla is not enthusiastic about the increasing migration into London over the last decades. She believes that 'this little island can't take any more'. Yet, on the other hand she enjoys the cosmopolitan mix of her residential street. Her neighbours include German-American, Cypriot-English, French-English, Tanzanian, and Italian-English couples. She also expresses excitement about having had a language student from Leipzig staying with her immediately after the fall of the Berlin Wall.

While it seems that Ulla welcomes the newcomer who relates to her cosmopolitan life style, she herself keeps in touch with certain parts of the globe. Having kept contact with Swedish relatives via her niece, who regularly comes to visit London, and having visited her brother occasionally in the States, travel activities increased in the late 1960s after her husband's death. Ulla now visits Sweden and America in turn each year, even though her frailty limits her mobility to a couple of streets within her neighbourhood. Over the years, if anything, the contact with her siblings has become stronger. Not too long ago Ulla had to nurse both her sister and brother on their death beds. Far from those transatlantic contacts ebbing away after the death of her brother and sister, special ties have developed between Streatham and Newfoundland, as Ulla continues to visit her friends, as she stresses, every other year.

Despite her cosmopolitan outlook and fairly mobile life style, Ulla feels settled in her quiet residential street in the Inner City suburb of Streatham, and is proud to have obtained the freehold of her parents' house. After her husband's death there were several offers from relatives in Sweden and friends in America to come over and permanently live there 'which of course', she says, 'I'm not doing'. She still can recall the Streatham of her childhood, and even though Streatham has radically changed since then, she is an active member of the Streatham Residents Association that tries to keep the locality's heritage alive. Ulla is also active in the local church, and was involved with the citizens' negotiations over the future of 'Pratt's' department store with the local council. She is good friends with the neighbours around her, in particular with a Swedish-American couple from across the road. Despite all these local engagements, Streatham cannot claim her sole sense of belonging. Like she says, 'I belong to a lot of things'. For instance, apart from being active in the local church, she is also a member of the Anglo-Swedish Church in Central London, though she only goes there once in a while, for instance, for the annual Christmas bazaar. Moreover, Ulla makes clear that her sense of belonging is related to the whole of London rather than just Streatham. Streatham to her feels quite 'insular' at times, whereas she has 'always felt more international'. When pressed about what it is that generates this sense of belonging, Ulla refers to London's 'atmosphere', the 'atmosphere of London's theatre world'. It was this attraction to London's spirit more than anything else that made her come back after she tried to live for almost a year with her sister in Newfoundland.

Harold Ford – the 'global business man'

Harold, a citizen of the United States from Texas, has been living in London for about nine years. Now in his fifties, he works for a transnational chemical corporation, with its headquarters in the States, and its European head office in London. Harold and his family have lived in Kingsbury, North West London, for about seven years now. Before that, they lived for a couple of years in the St John's Wood area, traditionally favoured by American expatriates. His son goes to the American School in London, while his daughter has returned to Texas to attend university. Harold has retained his house in Texas and the family returns there almost every year for the longer holidays, and Harold dreams of retiring there raising cattle.

Harold and his family's move to London is purely related to work requirements. Before coming to London Harold worked in the Middle East and South Africa. He then became promoted as a, in his words, 'firefighter' for the European section of the corporation. This implies on-the-spot decisions about transactions and problem-solving, thereby keeping a 'transnational overview over the individually operating profit centres'. His job means that he is travelling virtually all the time all over Europe, and occasionally back to the States. His travel patterns reflect the changing economic landscape of Europe, now taking in places like Halle and Leipzig, centres of the former Eastern German chemical industry. Harold talks about his movements between places like Brussels, Frankfurt, Barcelona, Zurich, Munich and Rotterdam as if they were part of a local public transport network. The pattern of a normal week would see him flying out on Monday, and returning late on Friday. During the week, Harold is somewhere in Europe, flying, sometimes driving, between destinations. Accordingly, he relies heavily on standardized facilities such as airports, hotels, and fast-food outlets. But far from these settings being all the same to him, Harold says he has developed a 'strong attachment' to hotels and certain restau-rants in each of those cities, prefers 'McDonalds' to 'Burger King', and feels safer at Frankfurt airport than at Brussels airport. Together with his own 'little routines', like keeping change in different currencies sufficient for a taxi ride, he calls this pattern, which he developed over the years of travelling, his 'travel sense'.

His residential neighbourhood seems to provide the counterbalance to Harold's restless mobility during the week. He talks affectionately about 'the village' when referring to his suburban neighbourhood in Kingsbury. While leaving the mundane neighbourhood activities during the week to his wife, he talks about the more symbolic festivities like the annual 'street festival' and the 'village garden contest', for which Harold, as a keen weekend gardener, has great enthusiasm. He knows one of his neighbours 'quite well', as he cuts her shrubs occasionally, but otherwise the 'village' seems to be more of a cosy refuge from work stress than a place for intense socializing. Instead, social contacts developed 'mainly through the children' and their involvement in the American School, and also through the weekly attendance of the Lutheran Church in the City of London, a meeting place for many London expatriates. Here he finds people with the same interests and problems. Harold enjoys going there on a Sunday morning for reasons of socializing more than worshipping, because, as he emphasizes, 'you are kind of looking for someone like you, somebody that has the same problems that need to be discussed'. His company setting does not provide these necessary social contacts, as, according to Harold, every one is on the move all the time, a 'pigeonhole' office where people leave messages for each other, but hardly see each other a lot.

Considering the circumstances through which he came to London in the first place, and taking into account his absence during the week, it does not come as a surprise when Harold describes his relationship with London as a fairly distanced one. London for him is 'purely residential' and 'not a city of choice'. Harold draws a distinction between 'my village' and 'London', which he associates with 'traffic jams, lack of parking space' and 'having to muddle through the crowd'. Harold is aware of the fact that, unlike many of his visitors from the States, he does not take any interest in London's 'theatre land', knowing however, that he will have to see a few musicals before retirement in Texas, as 'everybody would want to know if you saw *Phantom of the Opera*, *Miss Saigon*, *Cats*, or whatever'. On the other hand, he takes some pride in showing visitors from the States around the East End and the new Docklands, contributing to what he calls the 'myth of London' that, in his experience, is still powerful, particularly for people from 'small town America'. His daughter also confronts him with this 'myth', as she marvels about London's greatness with regard to culture and entertainment each time she returns from university in Texas. Despite having travelled the world, and having lived in different parts of the world, Harold's outlook is certainly not cosmopolitan. On his travels he tends 'to not make any contacts' in order to concentrate on work, while London's multicultural variety is something he rather avoids. When, for example, the Notting Hill Carnival is on, Harold would rather 'go for a picnic if the weather is nice'. Accordingly, he hopes to keep his 'village' unspoiled by all that, believing that this is something he shares with others, as in his road 'there has just been some silent agreement not to sell to Indians and Pakistanis'. Harold has no intention to leave before his retirement, since, after all, London is an 'easy city', a good base for travelling, and where everyone speaks English.

Nicos Calacuri – the 'local entrepreneur'

Nicos arrived in London at the age of 11, with his father fleeing the Greek–Turkish conflict in Cyprus in 1974. The sudden adjustment to London life he recalls as a harsh experience, coming from a rural background and suddenly having to 'run around all the time'. They 'struggled a great deal at first'. But after two years Nicos' father was able to start his first business, a Bed and Breakfast place in Paddington. By his own admission, he never really adjusted to the rat-race of London life, though he has tried very hard. With his father, Nicos went through a succession of hotel and catering businesses, working his way up to lower management positions in London's catering industry. In the early 1990s he finally set up his own little Greek restaurant in Streatham, only for it to close a year later in the context of a general recession. He is now back in catering management, 'working for other people again' as he says disapprovingly. This life trajectory has obviously influenced his views about London, which he describes as 'a jungle in which the weak are just left behind'. Central London and the West End he associates exclusively with work.

Since his life has been determined by the 24-hour rhythm of London's catering industry, he has not had much chance to enjoy the bright side of London life, such as London's theatre land. He would much rather spend the few hours of spare time he gets enjoying local entertainment, going to a Bingo Hall or one of the two cinemas in Streatham High Street. At home he catches up with all available news on Cyprus via satellite television, which he also uses in order to follow European football. Occasionally he drives up to Nottingham to treat himself to a home game of his favourite football team, Nottingham Forest FC.

Though he owns his own house in Streatham, where he lives with his English fiancée, he is disillusioned with life in London. Constantly comparing it with 'back home', an image that has recently been refreshed by a holiday in his grandparents' village, he argues that 'there is no real value to life here, it's about money, nothing else'. After nearly 25 years in the big city, something in Nicos wants to give up and return 'home'. However, there is this image of the 'golden streets of London', imposed on him by his relatives back home. Thus, returning without lots of money, the proof of 'having made it' in London, is not really an option. Nicos describes this image of London as an 'ideology', while the 'reality' of London life, according to him, is 'hard work, pollution, crime, and you are a prisoner in your own house'. On the other hand, he appreciates London as an 'organized city', with proper public transport, and where 'people are queuing for everything instead of pushing'. Nicos identifies a 'true' sense of belonging as something crucially missing in his London life. Though, or maybe because, in his job he has to deal with all kinds of people from different backgrounds, he does not believe in London's 'mixed society'. His scepticism is based mainly on negative experiences in business with 'con men' and 'posers'. Trusting someone for Nicos means knowing someone's background, 'knowing who they are'. Thus, Nicos believes 'that you're better off living in your own country, knowing your roots'. Not surprisingly, then, during the interview Nicos frequently refers to himself as a 'foreigner'.

However, despite the appeal of this ideal picture of everyone living in 'their own' communities, Nicos also comes across as an independent individual in search of the good life rather than in need of an ethnic sense of belonging. He never made any serious attempt to settle in one of the Greek-Cypriot clusters in Hackney and Harringay. In fact, at the age of 22 he turned down an arranged marriage, in favour of enjoying single life, and he admits being happy with Vic, his English fiancée. And in more general terms, he assumes that possibly he would not really fit into a Greek-Cypriot way of life in terms of attitudes. For example, he likes 'a little flirt', something where he likes the more tolerant attitude of 'the English', while 'the Greeks are too hot-headed when it comes to their wives'. Also, most of his friends are 'foreigners' like him, by which he refers to people from France, Germany and Poland, where his best friend comes from. While earlier rejecting London's 'mixed society' on personal terms, in the context of his profession he does take pride in knowing how to relate to very different people. In his words, he knows the different ways of 'pleasing your customers' by 'knowing the different attitudes, the different styles and different ways of talking, of serving, and so on'. In this regard he thinks he has become 'streetwise' and has learned 'to be a real good actor'. Still, when talking about his neighbourhood in Streatham, his search for true and meaningful social relationships, like those 'back home', comes through once again. Nicos wants more than a neighbourhood that 'functions' in terms of crime-watch and exchange of emergency keys. For example, he cannot accept that a neighbour of 10 years just passes him by with a brief 'hello' when they bump into each other in Central London, or that his neighbours do not patronize his restaurant. Those are the occasions on which he feels that he still has a lot to learn and needs to develop a 'harsh attitude' in order to make it in London, if after all he intends to make it work for him.

Herbert Reading – the 'retired local'

Herbert, now in his early seventies, was born in the Elephant and Castle area of London. His father earned a living as a London bus driver, and his mother was a housewife; the family inhabited the first floor of a terraced house. After the Second World War the family was moved to a council estate in Morden, South London. After getting married, he lived for a while in Stamford Hill, North London, before moving into a village just outside the green belt of London.

From the age of 14 onwards Herbert worked his way up in a City of London post office, starting as a 'post office boy messenger', and retiring in 1985 as 'Deputy District Post Master'. He now enjoys his retirement in the village, pottering around and only occasionally going up to London's West End for a meal or to the opera. As he stresses several times, 'London has changed', for the worse as far as Herbert is concerned. He refers to the London of his childhood as a 'happy place'. In particular, he discusses the working-class neighbourhoods in Elephant and Castle as 'very much like a village' where 'everyone knew each other'. There was mutual help and shared customs, such as drawing the blinds when somebody had died, or collecting money if someone was called up for army service. Despite poverty and unemployment, according to Herbert, there was 'no big jealousy, and no crime to speak of'. In Herbert's opinion, today's London 'is no longer like that', mainly because London's population is now 'so transient' that 'there is not the same association with a particular area and the same road' any more. Overall, he believes he has noticed a 'decline of community spirit'. Herbert has been searching for a similar setting ever since he was forced to leave the Elephant and Castle neighbourhood during the Second World War, without ever really regaining 'this sort of get together spirit'. The estate in Morden was 'too anonymous'. But for a while at least, his mother-in-law's neighbourhood in Stamford Hill, where he moved to after he married, appeared to bring back the ideal world as 'there were all-white families down the road, and they were all like a street village'. But then 'coloured people started moving in', reason enough for others, including Herbert, to move out of that area.

In the commuter village he moved to in the 1960s, he feels he has 'regained a bit of what it was like in London in those days of my life'. His social life evolves around the local community. Herbert says he and his wife 'got involved in things in the village', with him being an elected member of the parish council, and his wife having run the village play group for the last 20 years'. He enjoys the fact he's known in the village, and likes meeting familiar faces when he goes to the local pub or just for a walk in the village. Herbert feels fairly happy in this environment. Apart from a brief excursion to Canada for pilot training during the Second World War, he has never really felt an urge for travel or exploring. For him this is not just due to a lack of opportunity in his early years, but also a question of 'getting your priorities right'. And in this regard he always felt that 'the roof over your head' must be 'first priority', not 'fancy holidays'.

The cosy village life also provides some shelter from the increasing stream of immigrants into London. Herbert makes a clear-cut distinction between the 'indigenous population', meaning English people, and 'strangers', under which he counts not just 'coloured people' but also Australians and Germans, in fact everyone who is not English. Herbert insists on not being racist, and has nothing against those people individually, but claims that 'we will sink under the weight of them all'. Again, he refers back to the good old days before the war, when there were 'foreigners' in London, but not in such numbers. He explicitly mentions the Italians and the Jews as people who would 'have their own business', would be 'quiet', and who 'kept themselves to themselves', while today's 'strangers' Herbert associates with crime and loud music. The only scenario Herbert would find remotely acceptable is one in which 'numbers are kept within reason, and the people live according to your rules, speak your language, and work honestly'. Herbert clearly sees that the foreigners have not spread evenly, but are concentrated in cities like London. The only coloured person in his village is the shopkeeper, 'and he doesn't even live here', and Herbert wants things to stay that way. However, he is aware of the fact that the rural tranquillity he enjoys in the immediate neighbourhood of the metropolis lives on borrowed time. The village only recently fought off a planning application for a Buddhist Temple to be located in the open fields next to the village. But just a few miles down the road, in the commuter town of Stevenage, where his daughter lives, Herbert sees signs of a 'multicultural society'. Thus, Herbert is prepared to enjoy the shelter of his village life for as long as it lasts, watching the nearby metropolis fading ever further away from the London of his childhood and youth.

Ira Braithwaite – the 'metropolitan'

Ira's parents were amongst the staff recruited in the Caribbean for London's public services during the 1950s and 1960s. Their stay in London was meant to be temporary, in order to support family members at home in Barbados. But they happened to stay, as Ira 'came along', 'the first Londoner in the family', as she proudly states. Now in her late twenties and studying for a degree, as well as working part-time in a book shop, Ira has 'never really lived outside London'. She describes London several times as 'my home', and wonders if she 'could ever live outside London'.

However, this enthusiastic view of London was not something given to her by birth. Quite the opposite; Ira had to work hard to make the metropolis her home. Growing up in a council estate of 2,000 people in the Battersea area of London, Ira went to the estate-run play scheme, while her mother worked in the estate co-op shop. Later, she went to the same school as everyone on the estate, and hung out with the estate gang after school, going to each other's parties, and dating 'the boy from number 169'. Ira felt that there must be more to life than that 'totally entrenched' estate life, and consequently left home at 16 at the earliest opportunity, upsetting her friends, parents and sister. Still, Ira did not regret her decision at the time, and never has since. She knew that estate life was 'holding me back from being me'. Desperate 'to learn about myself' and 'yearning to meet other people', Ira moved into a shared house. This turned out to be a sudden introduction into metropolitan life, as Ira shared with someone from Spain, a 'guy from up north' and a gay couple. During the following years Ira moved house almost every year, always living in shared houses or flats. Ira stresses that in those years she learned how to share an environment without invading someone else's life. She also learned to live with other people's prejudice and to tolerate different points of view. It helped her 'to open my avenues to people' and to embrace metropolitan variety whole-heartedly. She says she realizes how much she has changed over the last 10 years whenever she returns to her family on the estate, experiencing the 'great gulf' that has opened up in particular between her and her sister.

Over the years Ira has lived in such different places as New Cross, Putney, and Islington. While on the one hand Ira stresses that she has 'not formed any particular bonds with those places', she likes each of them for their particular 'atmosphere' and the 'people' who create that atmosphere. It is for these reasons that she now feels rather settled in Brixton. She says she can 'hear and feel Brixton' as soon as she climbs the stairs of Brixton tube station. At the same time she is reluctant to call Brixton 'home' as she 'is not really part of the local Caribbean community'. Home is for her where she 'can have a bath without having to hurry, having music and a drink with it'. Despite the fact that she has no particular social contacts with any neighbours, Ira regards the street she lives in as 'home'. In this respect she describes 'a feeling of coming home to something familiar' and says she enjoys 'seeing the same people, even if I don't talk to them'. Ira's social contacts are 'scattered all over London', while she has not made any friends in Brixton yet. Socializing is therefore something she associates almost entirely with Central London, choosing places that are convenient for both her and her friends. Alternatively, she is quite happy to go about town on her own. Ira knows the metropolis and its rhythms to such an extent that she feels in control of them rather than exposed to them. Depending on the 'vibe' she feels, she might go and sit in a West End pub and watch American tourists leaving the theatres, or she might look for a place where she can be anonymous, just listening to her type of music, while on a Friday night she might go to a single person's type of place in order to 'give myself the opportunity to meet other [single] people'. Ira also likes the variety of London's street markets, and events like the Notting Hill Carnival or New Year's Eve at Trafalgar Square are highlights of her 'London season'. During those events, she says, 'you feel part of it all, you feel young and vital'.

Though Ira realizes that there is 'room for disharmony', she thinks London is unique in the way it has 'accommodated its foreigners'. In comparison, Ira experienced New York as quite a segregating place. London allows her 'to have a touch of everything in my life', while 'over there', visiting a cousin in Brooklyn, she had to absorb more 'black culture' than she wanted to. Ira has also been to visit some of the European cities. In particular she liked Rome and the Italian life style, but found it funny that people 'wanted me to be from Nairobi or something'. She has loose contact with the extended family in Barbados, meaning about six telephone calls a year and three holidays so far. Barbados for Ira means lying on the beach and sipping a 'Barcadi-Coke', but by no means does she nurture the dream of 'going home' one day. She says she would miss the crowded London buses and the atmosphere of Brixton. So, while her parents still hope for her to be taken home by 'Mr Barbados', Ira has no intention of ever leaving the metropolis, the London she calls 'my home'.

Barbara van der Velde – the 'expatriate'

Barbara, in her late fifties, is a missionary with the Methodist Church. She is originally from Nashville, Tennessee, but has lived substantial parts of her life in Mexico and the Philippines, from where she came more or less directly to London. She came to London when she applied to WACC (the World Association of Christian Communication) to be part of the organizing team of a congress to be held in Bangkok. Barbara is engaged in the communication preceding the conference. She has to mail out details to 400 delegates all over the world, and respond to inquiries, mainly via e-mail. As the conference is on 'women and media', part of her job is to keep in contact with two participating bodies, the ISIS (a non-governmental women's organization providing information and communication services) based in Manila, and the IWTC (International Women's Tribune Centre) based in New York. In the little office in the Vauxhall area of London where the organizing committee is located, she works with colleagues from France, Switzerland, the Netherlands, the Philippines, and Portugal.

37

By the time we talked, Barbara had been living in London for six months, a quarter of the way through a planned stay of no more than two years. She is thrilled to have the opportunity of working in London, a city she has always wanted to visit. When she arrived she was full of anticipation based on media images from the Houses of Parliament, the Royal Family, and the Wimbledon Tennis Tournament, but also based on the stories of 'returnees' who told her London 'is a wonderful place, you'll find it very nice'. She thinks it is 'really fun' to now finally experience 'something we've been admiring from a distance'. Accordingly, she has tried to make good use of her spare time during the first month, and has even managed to get a ticket for the Wimbledon Tennis Tournament. But there are other things which 'amazed' her about London. Barbara still has to take in the 'sheer diversity of people here'. While she thought 'you are going to be with British people', she finds herself listening to not just many different English accents and dialects, but also French, German and Spanish when travelling to work on the London underground. She also mentions the 'many mixed marriages' amongst her neighbours in Streatham, and amongst the people she meets on Sundays when going to Church, as something 'which is, I think, less common where I come from'. In Nashville, she says, 'you would see a lot more separation of people by the areas they live in'. In general, she has formed the impression that Londoners 'are more willing to accept differences'. For a newcomer like her, she says, this is a 'very comforting feeling'. Despite being nearly 60 years old, Barbara says that she is 'growing with this experience', becoming tuned in to the fine distinctions of culture, being more aware of her use of (American) English, and in general 'appreciating the people I met'.

When Barbara arrived in London, she was equipped with a couple of contact addresses in London and surroundings, given to her by friends in the States and elsewhere. Both at work and in the Lutheran Church in the City of London she has found people that she likes 'to associate with'. This is the case as these are social settings where she is amongst other 'expats' like her with similar types of interests and problems. But more specifically, she tends to spend most of her spare time with two women from work 'who also have a history with the Philippines', and with a Filipino couple whom she met at the Lutheran Church. With these people she explores London's cultural variety, including visits to the Tate Gallery and attending festivals of the Filipino community. Perhaps not surprisingly then, Barbara states that she 'feels at home with them' (the Filipinos), mainly because of the time she has spent in the Philippines. The same feeling of familiarity and home does not (yet) apply to her residential neighbourhood in Streatham, where she rents a room in the house of another American expatriate family. First, because the place of residence was pre-arranged, not a choice, but mainly because for Barbara 'home' is not primarily related to places but is 'where I feel comfortable with people'. Still, she tries to make her rented room 'my little world'. She has brought pictures, books, newspaper clippings, a little radio that accompanied her to other places, and letters from her children – 'things which are meaningful to me and for other people'.

In order for her to feel settled in a place she says she needs to keep in contact with friends and family in the States, Mexico, and the Philippines. She is in contact with her son at least every other day via e-mail at work, while her daughter phones regularly, and they exchange 'snail mail'. However, she mentions that she is restricted in her communications, as for the next two years she will have to live on a fairly tight budget. That means, for example, not making expensive phone calls more often than necessary. Barbara stays in touch with friends in the Philippines via letters, and she still receives the newsletter of her old parish back in Nashville. Not surprisingly, Barbara refuses to be pinned down as to where 'home' is for her. Though it seems to become 'perhaps more London now', she still feels strong attachment to the States and the Philippines. There are rather technical ties to the states, such as a house she still owns, or a pension from her divorced husband. And there is a friend who keeps a 'safe-deposit box' for her, including amongst other things, her will. As for the more emotional attachment to the Philippines, she has already applied for posts there and in Hong Kong for the time after London.

Rolf Grauer – the 'reluctant explorer'

Rolf, from Bavaria and in his early thirties, came to London rather by accident. His girlfriend got a job in London, and he was more or less accompanying her. While she was preparing the base for the new life in London, Rolf was supposed to pack up things back home in Germany before joining her. They found a flat through the recommendation of a friend in London, whom they already knew from a music workshop in Germany. He said that in this way he had had a 'rather soft start' in London. In other ways the move to London turned out to be more challenging. He had a tough time finding employment in his profession, computer programming, finally finding work with a small company on the outskirts of Greater London. In their spare time Rolf and his girlfriend started to explore London's rich classical music scene. Although they had some idea about it, they felt pretty 'overwhelmed' by the high quality options to choose from. Rolf believes that no city in Germany can compete in this regard with London, be it Berlin, Hamburg or Munich. The same applies to the many talented amateur choirs, one of which Rolf decided to join. He generalizes that 'London has more to offer than any city in Germany'. What particularly impressed him is 'to live in such a conglomerate of nationalities', again something, he believes, that seeks its equivalent in Germany. He thinks that Munich is rather provincial in this regard. He recalls from his youth that 'seeing a negro was a world sensation', and he still associates them with the 'Amis' (American Forces in Germany) rather than with everyday life. In London he finds himself living and 'singing alongside them', in the truest meaning of the word. At work, he estimates, half of the 36 employees are 'foreigners'. Again, he draws the comparison between working with colleagues from France, Bulgaria, Australia, Ireland and India, while at home the one Argentine that his company employed was 'considered somewhat exotic'. The only thing he does not really like about London is its hugeness. More than he would like to, he finds himself travelling across the metropolis between work, home, and his singing. One drawback of that is that he nowadays 'tends to rely on fast-food'. As a necessary evil he prefers McDonalds, 'because it's fast and you know what you get between you teeth', at least in comparison to the plethora of mobile burger grills with their 'tasteless floppy Hot Dogs'.

While initially Rolf threw himself into work and his singing hobby in order to 'survive' the three years his girlfriend intended to stay, the metropolis has had a rather lasting impact on him, which has changed this negative overall perspective. The more he opened himself up to the metropolitan variety, with its variety of life styles, the more he felt relaxed 'and not alone' with something he had not wanted to realize or confront in Germany, namely his homosexuality. He says that London helped him to 'come out' and 'learn to stand on [his] own two feet in this regard'. He split amicably with his girlfriend and moved to another place of his own in Forest Hill, South East London. He is now exploring London's gay world, stressing at the same time that London is a place that makes sure that 'you can't really get entrenched in your own little world'. Most of Rolf's social contacts are now in London's gay scene and choir scene, both network of acquaintances and friends partly overlapping, as he now also sings for the 'Pink Singers'. But Rolf also likes to just hang out in one of the pubs around busy Covent Garden or Leicester Square, 'just watching the different types of people walking by'.

When talking, he sometimes refers amicably to 'my London', expressing a certain pride he feels when showing friends and relatives around the city he finally conquered. Rolf says his new attitude towards London also came home to him when last Christmas, back home on holiday in Germany, he felt 'gosh, another 10 days; I wish I could catch a flight tomorrow'. He also decided to 'slow down a bit' in order to appreciate life in London even more. So does he feel 'at home' in London then? That depends. He refers to himself as a 'foreigner', believing that his not quite perfect English would make that 'bloody obvious to anyone you talk to more than two sentences'. As far as his gay sexuality is concerned, he feels that he has finally 'come home' in London. This feeling of relief and achievement also seems to transfer to his actual place of residence, a one bedroom flat in a modern style house. He says he likes it much more than the one he shared with his girlfriend. His new place is his 'home' in that it symbolizes to him the 'freedom of a new-found life'. And he is not prepared to compromise that freedom. He say that he keeps contact with neighbours in the house to a minimum, and as far as the customary exchange of keys is concerned, Rolf would 'rather rely on a key cutting service' in case he gets locked out, or 'wait for the landlady to come from Wales'. Otherwise, he doesn't really know his neighbourhood, as most of the time he is out working, singing, or enjoying London's West End. Though he is a bit concerned about losing track of professional developments in Germany, Rolf says that 'at the moment I can't see myself going back to Germany'. For the moment he is quite happy not having to make a decision between London or Germany, with friends and his mother visiting him from time to time, and him going over twice a year for longer holidays. His car insurance and lots of other technical things are also being dealt with in Germany by his mother. So he says he can imagine 'another 5 or 10 years here . . .'. Should he return, he is confident that he will then take his 'new self' with him. Thus, after nearly two years in London, and with a certain sense of irony to it, his girlfriend is leaving London prematurely due to a job offer in Germany, while Rolf, who was initially reluctant to come to London, will stay on.

4

THE UPROOTING OF
MILIEUX

> The city has always been an embodiment of hope and a source of
> festering guilt: a dream pursued, and found vain, wanting, and
> destructive.
>
> Raban (1990: 17)

Behind each of the biographies sketched out in the previous chapter there lies a story about an individual attempting to bring their milieu in tune with the pulsating rhythm of the world city. In their more or less successful attempts to take on both the challenges thrown at them and the options given to them by metropolitan life, each of these milieux contributes to what Raban calls the 'melodrama' staged by the city (1990: 28ff.).

The city is a place where we meet the 'other' and encounter the social practices and beliefs that come with the 'other'. These encounters have the potential to challenge our own routines and preconceptions about the 'good life'. A whole tradition of social theorists, from Spengler and Simmel to Sennett and Mumford have stressed the potentially 'uprooting' impact that city life has on our milieux. The latter, for example, maintains that the city is 'designed to make man at home with his deeper self and his larger self', demanding continuous 'self-actualization' (Mumford 1991: 652f.). Kiwitz, in taking Mumford's argument a step further towards a phenomenological analysis of the urban environment, emphasizes the city as a setting that escapes systematization and encourages individuality. He highlights the city's unique way of accommodating difference at many levels of everyday life, allowing for 'creative impulse' and encouraging, or perhaps rather demanding, the 'release of human vitality' that remains discouraged or even suppressed in other settings. Kiwitz subsequently aptly defines the city as the 'affirmative territory of the life world' (1986: 149ff.).

While this could possibly apply to some extent to any city, it certainly holds true for a world city like London. This chapter illustrates this 'uprooting' potential of London life by taking a closer look at three of the biographies introduced in chapter

3. Each of these tells a different story about the interplay between the individual's milieu and the sociocultural environment of the globalized world city, London. Contrasting the respective initiating impulses, we can argue for a differentiation between external and internal uprooting. While the notion of *external uprooting* stresses the challenging impact a complex sociocultural environment can have on a rather closed milieu, *internal uprooting* in turn highlights the milieu's vitality in breaking through the constraints of a given social environment. In order to counterbalance the idea that uprooting equals a liberating impact that benefits the individual's milieu, the notion of *disrupting* shall then serve to highlight the fact that there is also potential for a structural misfit between milieu and social environment.

Uprooting, as described and analysed in this chapter, is certainly akin to Giddens' concept of 'disembedding' (1994: 21, 53). Yet, while disembedding stresses the spatio-temporal aspects in the transformation of the 'situatedness' of milieux, uprooting puts emphasis on the fact that this transformation also has an affective or aesthetic dimension. The milieu is not just a configuration of spatio-temporal routines but refers at the same time to the individual's general affective disposition. Uprooting, then, also implies a challenge to the individual's open-mindedness. The point here is that these two aspects, spatial and affective uprooting, are most probably complementary. But, as we will see, the 'lifting out' of a milieu from its local contexts does not necessarily go hand in hand with a 'broadening of horizons'. The chapter thus serves to indicate that disembedding processes are not necessarily large scale, but can also manifest themselves in quite subtle and, at times, personalized forms.

External uprooting – coming out

As we have seen, Rolf, who was introduced as the 'reluctant explorer', came to London not out of choice, but because his girlfriend had a three-year work contract there. He followed reluctantly, regarding the move from his home environment in Bavaria to the big city as something of a necessary evil. As he saw it, if he wanted to maintain the relationship with his girlfriend, then he had to show an interest in moving to London. In other words, London became of relevance to him because someone else imposed the issue on him. With Schütz, one could say that London becomes an 'imposed relevance' in Rolf's milieu, meaning that the task of relocating and reorganizing his life in London did not derive from the intrinsic trajectory of his own life plans (cf. Schütz 1970: 178). Not surprisingly, then, Rolf felt not in the right 'disposition' to move, and the whole project of moving and living in London remained somewhat external to him. He initially considered the three years as literally something of an 'interruption' of his own milieu, which he hopes to re-establish after returning to his familiar home environment after the three years had elapsed. Accordingly, the coping strategy Rolf originally pursued in order to maintain his milieu in the new environment is to escape through a kind of time warp. By working long hours in his computer job he hopes that the three years in London will pass rather quickly:

> I thought, OK, if I know that it's really only for two or three years, and then I can go back, I can throw myself completely into work. So I started working like mad, 10 hours a day . . . and so the time til April passed very quickly.

However, his interest in classical music and church choirs was something he shared with his girlfriend, and that subsequently opened his milieu towards the cultural and social environment of London. Enthused not just by the options for passive enjoyment but also the opportunities for active participation, Rolf started to ease his resentment towards London:

> I mean, this is exactly the little bit extra that London has to offer, I mean for people actually to participate, there is much more variety here than in any city in Germany . . .

From here, one thing led to the other. Once he had gained access to London's network of amateur choirs, and soon afterwards started singing for the prestigious 'London Bach Singers', he started to encounter and enjoy more of London's entertainment and night life. With his girlfriend becoming more ingrained in her work, Rolf became more and more of a 'flâneur':

> Sometimes it was really, like, going out each evening, coming home at 6, getting changed, tube ticket, back into town, choir rehearsal, and then afterwards I often just stayed on for a couple of beers on my own. Doesn't bother me, I used to do that back home. I actually like that, just standing in a pub in Covent Garden and watching people walk by, just watching all these different characters in London . . .

It was during this normal course of events that the actual 'uprooting' of his milieu began. Rolf increasingly frequented what turned out to be gay bars and pubs, initially because of their more often than not extended business hours – or so he rationalized this to himself. But it became clear to him that there was some aspect of his self that had been suppressed for a long time. Frequenting these gay pubs encouraged him to confess to his gay sexuality which he had hidden from his family back home, from his girlfriend, and partly from himself. He clearly identifies the importance of London's cosmopolitan everyday life as crucial for what we could call the breaking through of suppressed sexual relevances in his milieu:

I think it was this context which helped. I mean, this really gave me the courage and power to actually do it. I mean this is great, to just go into a gay pub, and just stand there, just to be normal, to know that you are not alone, that you are rather crushed by people like you . . . And this really gave me the power, still gives me the power to go out on my own and actually be who I am. And this simply wasn't the case back home in Germany. I don't know, it was just different. In this sense I'm pretty sure that London . . . I don't know if the same would have happened in Germany.

The breaking through of formerly suppressed relevances subsequently restructured Rolf's milieu in London quite dramatically. He separated from his girlfriend, moved to another place and started to actively socialize with the like-minded. With the change in his disposition, formerly established relevances of his milieu also became transformed, like for instance his singing hobby as he joined the 'Pink Singers'.

What is important to note here is that the 'uprooting' of Rolf's milieu in terms of his general disposition, also finds a spatial expression. Moving away from the place he shared with his girlfriend is an attempt to bring distance between his 'old' and 'new' self. The uprooting in his mind is underlined by the uprooting from a certain locale and the experiences attached to it. His new place is Rolf's first real 'home', the place with which he associates his newly found 'freedom':

Actually, I must say I like this place much more (than the place we had before) . . . because this is where I've found my freedom, I just love this place.

Although initially induced by his suppressed homosexuality, the 'uprooting' of Rolf's milieu develops wider implications. He feels that he has won freedom in a more general sense. Having just set his milieu free from the social norms and expectations of a certain social environment, Rolf is not prepared to exchange old constraints for new ones. Rather intentionally, he chose as his new place of residence a fairly anonymous housing co-op, meaning a minimum of neighbourly relations. He cherishes the anonymity of this particular local environment as much as the anonymity of London's everyday life in general.

I mean, the whole thing is a bit more anonymous in a city like London, I like that. It doesn't bother me at all, it actually gives me a lot of freedom. Just take the fact that I'm not tied to any neighbours here, you know, where you would have to invite them over to your flat because they have invited you over to theirs, and all that; I don't want that stress any more. I really want to enjoy my freedom for a while, after having been together with Rita for eight years, simply to kick over the traces a bit, you know, no more pressure.

One can detect a certain anxiousness in Rolf's statement about being 're-rooted'. And this applies not just with regard to local social relationships, but extends towards the newly gained situatedness of his milieu in London's gay world. Rolf is well aware that his successful coming out owes much to London's general atmosphere of variety and tolerance. He is determined to not let his newly-found self be entrenched in a narrow-minded and exclusive gay world. Instead, he reflects a certain open-mindedness when he says:

> I think it is important that you don't get to fixated on this world, that you remain open towards 'straight' people, that you don't wrap up in your gay world. Overall, I think gay people are more sensitive towards differences, but I found that they too can have tunnel vision sometimes.

Apart from generating a more open-minded general disposition, and finding a significant spatio-local expression, the uprooting process has also transformed the temporal perspective of his milieu. Whereas before Rolf was wishing for the time to pass quickly by working long hours, he now wants to get the most out of his time in London:

> I said to myself 'slow down', otherwise time just passes you by, otherwise the three years are over and you are back in Germany, and then afterwards you ask yourself, what did I do all that time, just work? You must be mad, life has just passed you by.

While he plays with the idea of extending his stay in London beyond the initially envisaged three years, Rolf has actually no doubts about the finality of the changes in his milieu. He seems to be pretty clear about the fact that the 'uprooting' of his milieu took place most importantly in his heart and mind, and will not be lost should he have to move on to another environment:

> I think London has changed me, at least in this one regard, where I have learned to actually stand on my own two feet, I have learned to move freely in my world, and that's something I will take with me from here to Germany.

The story of the 'uprooting' of Rolf's milieu in London told so far, reveals the need for a clear analytical differentiation between *milieu*, referring to the individual's unique life-world, on the one hand, and *environment*, as related to the objectively

46

existing external social and material surroundings, on the other. For this purpose an excursion into Scheler's theory of the milieu should prove useful. We have to recall that for Scheler 'milieu' in the first instance is not the same as material surroundings, but a 'value-world as effectively experienced in practise'. In other words, the milieu serves as an 'alphabet' through which the individual reads the environment as a meaningful and practically structured configuration. He goes on to draw the important distinction between *structure of the milieu*, meaning the individual's value-related disposition towards the world, and 'momentary' or *practical milieu*, referring to the materialization or realization of this disposition in practical engagement with the respective environment at hand (1973: 139ff.). In a more metaphorical sense, one can say that the milieu serves as a relatively stable filter through which the individual experiences different environments. Conversely, a given environment becomes a practical milieu to the extent to that it reflects the individual's disposition.

What follows from this is that milieu and environment 'always fit one another in a certain sense' (ibid: 134). As we can see from Rolf's initial engagement with London, a basic practical milieu had to be established, however reluctant he felt towards his new environment. The arrangement of a practical milieu, however, can mean that milieu and environment can either almost coincide but can also be more or less in conflict (ibid.). This means that a practical milieu will almost certainly carry some unexpressed, hidden or suppressed relevances. Moving on to another environment will then potentially change the balance between expressed realized and suppressed relevances. In Rolf's case we obviously talk about such suppressed relevances, eventually liberated by an unintended change of his social environment at hand. The reluctant move from small-town Bavaria to the metropolitan environment of London turned out to have this liberating influence regarding his sexual preferences, subsequently leading to a radical restructuring of his practical milieu.

The point to stress here is that a given social environment can stimulate or constrain the individual's milieu. Before the move to London the suppressed relevances could not become a generating force in his practical milieu back in Germany. Whether in the long run the same would have happened in his home town environment is a purely speculative question. The fact is that the pluralistic and tolerant tone of London's everyday life encouraged the, in this case rather literal, 'coming out' of sexual relevances that were part of Rolf's milieu anyway, but never before actually materialized in his practical milieu. However, not to be mistaken, the emphasis here is not so much on Rolf's sexuality as such, but, as his case illustrates, a successful uprooting of someone's milieu can be undoubtedly attributed to the sociocultural make-up of London. While anonymity and tolerance provided the appropriate frame, the option to actually encounter the suppressed relevances of his milieu as a lived normality, plus the chance to share his new milieu with the like-minded, certainly worked as a catalyst for Rolf's 'coming out'. The immediate presence of different social and cultural worlds, the high degree of normality in the coexistence of virtually all imaginable life styles within a few square miles, and the stimulating atmosphere that accompanies this concentration of variety and difference, make the microglobalized world city a place of potential uprooting and setting free of individual life trajectories.

It is important to also note that while a certain sociocultural environment can well encourage and stimulate a milieu, it can not create a milieu. Rolf at least sensed his homosexuality before he came to London and became involved in its gay world. This is why it seems to be appropriate to talk about 'suppressed' rather than new relevances that are set free, but not created in the context of Rolf's milieu. It is important at this point to recall that the practical milieu, in Scheler's sense, displays the 'dynamic relationship' between the individual's milieu and the environment at hand. Far from being a pre-stabilized harmony, it is a relationship of 'effecting and suffering', 'winning and succumbing' (1973: 137).

This point deserves further discussion, as the analysis of Rolf's story, dominated by the challenging contrast between his home town environment and London, puts another issue on the agenda. We could ask, what about milieux that have not experienced the challenging impact of moving from a rather philistine small town environment to a cosmopolitan city? In other words, what about people who have lived all their lives in a place like London, who, on the surface it would seem, miss out on the challenging and stimulating impact that London has had on a newcomer like Rolf? If there is something like uprooting processes going on, what is the driving force behind it? To pursue these and related questions it seems appropriate to look at the case of Ira, a Londoner by birth, and of Caribbean descent.

Internal uprooting – leaving the estate

Ira's biography, in contrast to Rolf's, provides an example of the 'dynamic' between milieu and the sociocultural environment being driven by the individual's determination to construct 'a life of one's own' (Beck 1992), rather than being induced by a stimulating change in the environment at hand.

Ira grew up on a large council estate in Battersea. All social contacts and relationships centred around the estate community of 2,000 people. While her sister and her parents were, and still are, happily involved in this localized community, Ira felt the urge to escape this life at the earliest opportunity. She claims that the incentive to uproot herself from estate life and her family came from inside her. Instead of posing a positive challenge, her immediate social environment only provided a negative contrast to what Ira expected from life. While her family was 'totally entrenched in estate life', as Ira expresses it, she always felt that there must be something more to life than 'going to the estate-run play scheme, going to the estate-related co-op shops, going with the estate gang to the cinema, going to the same school, going to each other's parties, and dating the boy from number 169 . . .'. She summarizes her feelings shortly before her escape as follows:

> I sort of hated Battersea when I was 16. I sort of hated Battersea, and I sort of hated living on the estate. I sort of hated my parents, I sort of hated the flat we lived in, you know, I always felt like all that was holding me back from being me.

Her decision to finally leave the estate came rather abruptly, leaving her parents and friends 'under a big cloud', and yet, somehow, she felt good about it. Ira clearly considered the estate environment as an environment that did not match her personality. Any attempt to be different turned out to be difficult in view of the constraints and expectations imposed on her by family and estate community alike. Finally moving into a housing co-operative where she did not know anyone, Ira felt freed from all these implicit and explicit codes of conduct that were never really hers.

> I felt good, because I think it's important to move away from your home territory, you know, so that you can be yourself a bit, you know, when you are in your home area you will always find that you do revert to what people there expect you to be like . . . And so I moved out, because I think I've always been of that kind, you know, I got to find myself, I got to know what I'm really like. And maybe I knew that I was gonna do things that were probably gonna annoy or make my parents feel ashamed, you know, because they are very homely.

It becomes obvious from her narrative that, in distinction to Rolf, Ira is not keen on finding out about, or releasing, a particular aspect of her milieu. She simply longs for variety in life, which the estate, though located in the middle of London, could not provide. Uprooting herself from the constraints of estate life, she now longs to just open her milieu, to 'open my avenues' as she calls it, towards the social and cultural variety of London. The fact that the first step towards that aim is a housing co-op just a few miles away, is not important. In its sociocultural make-up this setting proves just right.

> When I moved there I didn't know anyone in the house. But I grew up so much in that house . . . I was excited because, you know, there were all kinds of people, you know, I kind of love meeting people, and I love, like, opening my sort of avenues to people, and forming relationships. It was 10 different people living in that house, you know, there were all types of people there, and I really yearned for that.

However, there is no attempt on Ira's side to make this place her new home in terms of settling down. Her uprooting process is much more radical and gradual than Rolf's. Far from having found the social environment that once and for all matches her milieu, the move to the housing co-op is only a first step towards finding out about her milieu, – going on a journey to 'become a bit of an individual', as Ira terms it. And for her that seems to require not just open-mindedness, but deliberately exposing herself to new and altering social environments in order to once again put her own milieu to the test. Accordingly, Ira has moved on from the initial co-op several times, has lived all over London, always sharing with a variety of people.

> I've always lived in shared houses, I like that, I really love living in a shared environment, because it's like family, and yet it's not like family . . . I mean, you say things to them that surprise yourself, without knowing them directly, it's like a learning thing I guess. I definitely learned how to take people, you know, you can't learn things like that in a family, you can't learn about life like that, you kind of have to be out there and to experience it.

It appears that Ira consciously avoids re-rooting, and this does not just apply to her first housing co-op. It is more of an intuitive avoidance of a situation where milieu and environment are starting to develop a harmony. In this respect her uprooting is rather radical with regard to both local and social constraints. If things become too much 'like family' it is time to move on. Relatedly, locales and the relationships evolving around them are significant insofar as they contribute to the journey of discovery of 'what I'm really like', not, however, as shelters from the world.

> I think places, you know, some places have a significance for you if they kind of represent a changing period in your life, you know, or a discovery period in your life.

Ira is not quite sure where this journey of 'finding myself' will end, indeed, whether it will end. She is very much aware of the final consequences of her still ongoing process of uprooting, though. The move away from Battersea, from family and local community, turned out to be rather final. She knows that she had to leave her 'home' estate, though spiritually never really home to her, in order to make London her home, in order to be with the like-minded. Return is impossible.

> I couldn't live back in Battersea anymore. Once you moved out of home, it's different when you go back. There is no going back in a way, you feel

like a visitor, it's not like home, you are a visitor. Your home is where you're up, where your friends are, it's not gonna be where your parents are.

While the spatial side of her uprooting process is rather obvious in her moving house almost every other year, there is also a time dimension to this process. Nowadays visiting the estate in Battersea in a sense means visiting her own past, or the fairly predictable way she would have gone if she had stuck it out on the estate. 'I would have been like them', entrenched in estate life and hardly knowing about the rest of London. Especially in relation to her sister, for Ira this difference comes clearly to the fore.

My sister lived at home until January of this year, and I'm very close to my sister. But there is that thing, you know, there is that great gulf there, because, she has always lived at home, she has never moved out of Battersea ever, you know. So our lives have taken different paths, things have happened to me that she will never know about . . . It's funny, and it comes up sometimes if you are like watching TV or something, or I would crack a joke and she wouldn't get it, not because she is stupid, but because she is just not in that life style to know what I'm talking about.

Despite this detachment from her family and her tendency to move house every other year, Ira by no means feels like a 'homeless mind'. In fact, as we have seen, it is this quite strategic exposure to alternating social environments that helps her to become more aware of her own milieu, the things that drive her on. Just as she describes localities according to the significance they have had in the process of 'finding myself', there is also an immense sense of alertness regarding her social contacts. Rather than allowing London's social life to overwhelm her, she attempts to stay in control by choosing social settings according to her relevances, or as she says, 'according to the vibes I'm in'. With regard to Friday and Saturday night entertainment, for instance, she would know the places 'where you can go and find your type of people, your type of music'. Yet again, Ira is not searching for 'her local' in which she listens to the same music with the same people week in week out. With the relevances ('vibes') in her milieu changing, for instance after splitting up with her boyfriend, the venues she frequents in order to find the like-minded would change too.

I mean, you know, I'm in that single vibe at the moment, so I have been down to the West End a couple of times recently, because it's a focal point for giving myself the opportunity to meet other people.

Increasingly knowing how to play London's variety according to her own 'vibes', she has made London her 'home', rather than a particular neighbourhood or venue. What she appreciates about London is the way it allows people like her to uproot their milieux and stay that way, instead of being forced to re-root in some other setting that is strictly defined, be it by local, social or ethnic boundaries. Implicitly referring back to her New York experience, where she stayed with her cousin, she categorically states:

> As I said, I couldn't imagine living somewhere else, because I hate the idea of segregation, I don't like being placed in a situation where I feel like, you know, first and foremost you have to be black.

London, in Ira's account, is unique in that it 'can take care of so many people's needs', encouraging them 'to find their bit of London'. In this sense Ira regards London 'my home' in a way that can be shared with many other people like her, accommodating them without absorbing their milieux. It is London in its social and cultural diversity, rather than a particular locale or local community, that makes someone like Ira feel at home. For as an 'independent free spirited person', as she describes herself, she 'likes a bit of everything in (my) life'. However, as indicated before, the exposure to different people and life styles in Ira's case does not mean her own milieu getting lost in a plethora of possible styles. Rather the opposite, the encounter of difference is Ira's conscious way to become more and more her own self. She can thus confidently draw the line between her and someone who is a plaything of changing life style fashions in a vain search for identity: 'I wouldn't go into a particular style, I have my own style'.

Following Scheler's terminology at this point, we could say that Ira has found her 'destiny', though her self-imposed project of 'finding out who I really am' is still ongoing. Ira's story clearly features elements of what Scheler describes as the 'tragic conflict between destiny and milieu'. In keeping with his distinction between 'structure of the milieu' – that is, value-related disposition – and 'practical milieu', – the practical realization of this disposition in a given environment – Scheler goes on to differentiate between 'destiny' and 'fate' as the temporal dimension of this distinction. Insofar as the practical milieu is never entirely of the individual's own making, but instead displays the pre-given and accidental character of the social environment at hand, it has an element of fate to it. According to Scheler, it is then the individual's responsibility not to mistake this practical milieu for their destiny. Instead, the individual has to realize their destiny in an ongoing process of self-experience in practical life, constantly rebalancing self and practical milieu by following their own moral disposition. However, the individual's striving for a practical milieu that matches their 'moral tenor' can be distorted, as people can be fooled about their own inclinations by trying to live up to the expectations imposed on them by the 'moral environment' around them. In which case one could talk of a 'tragic milieu' in which the individual is withheld from self-realization (Scheler 1957: 349ff.).

Ira clearly felt that there was a conflict between her practical milieu and her self in that life on the council estate was 'holding me back from being me', as she puts it. Moreover, she realized that it was not just the practical milieu of this one particular estate, but an 'entrenched' estate culture in general that put spatial, social and ethnic constraints on her. Ira is aware that her leaving the 'home' estate, was not just uprooting herself from a particular locale, but from a 'moral environment' that expected her to stick it out on the estate and 'to wait for Mr Barbados to take me home', as she paraphrases her parents expectations. One reason for her to leave the estate was the feeling that she could no longer obey to the moral expectations of the estate community in general, and her parents in particular: 'I knew that I was gonna do things that was probably gonnna annoy or make my parents feel ashamed'.

Consequently, in order to 'get to know who I'm really like', there was no other option than to leave behind a whole structure of expectations and moral obligations for good. Otherwise 'you will always find that you do revert to what people there expect you to be like'. Moreover, for fear of jeopardizing her Socratic project – 'I got to find myself' – she subsequently tends to avoid any 'family like' situation, intentionally moving into new shared environments every other year. It is worth stressing here that, although on the outside it might have features of the 'homeless mind', Ira's Socratic project is not driven by chance. The confidence of being in charge of that continuous process of rebalancing her self and altering social environments in a succession of practical milieux is clearly expressed when she talks about '*giving myself the opportunity* to meet other people' (my emphasis).

This is the point that highlights the difference between Ira's 'internal uprooting' and Rolf's 'external uprooting'. While the crucial impulse for the 'uprooting' of Rolf's milieu concerning its suppressed sexual relevances came from the challenging and yet encouraging impact of a changed external environment, there was no such initial catalyst in Ira's case. There was also no lived reality of what was missing in her milieu, while Rolf had London's gay world at hand.

Compared to Rolf's quite focused 'uprooting' and subsequently rather final rebalancing of his milieu, Ira's 'uprooting' has no particular aim apart from gaining a certain open-mindedness, or 'opening my avenues to people', something that was denied her in the estate culture she grew up in. Attempting to find 'my own style' in a succession of 'practical milieux, Ira's story illustrates what Scheler calls the potential 'world-openness' of the human milieux, that is, the ability to 'distance' oneself from any given momentary milieu (1976: 30ff.). Ira's uprooting was initially entirely driven by such an intuitive distanciation. There was no experience of other life styles, no role model of an older sister, just this 'internal' feeling that the milieu she found herself in was 'holding me back from being me'. Ira's story, thus, also makes sense in the light of Scheler's assumption that to find and realize one's own milieu is essentially down to the individual's own effort rather than 'fate'. Recalling here that Ira's and her sister's lives 'have taken different paths' out of the same pre-given practical milieu only serves to highlight the importance of individual determination. Ira's story suggests that to find 'one's own milieu' is a life-long project rather something that comes naturally.

Global mobility (Rolf) and exposure to a microglobalized environment with its multiplicity of coexisting life styles (Rolf, and subsequently Ira), might make it increasingly likely for people to recognize the restrictions and the undiscovered potential of their milieux. Moreover, an environment such as London certainly encourages acts of liberation such as Rolf's, or ongoing projects of self-actualization such as Ira's.

What both Rolf's and Ira's stories have in common is their happy ending. 'Uprooting' in both cases means more than just leaving behind a certain local setting, but implies suppressed relevances of their respective milieux being set free. That this liberating element is not something which can necessarily be linked to the 'uprooting' of milieux shall be illustrated in the following by looking at Herbert's story, which confronts us with what could be described as a 'disrupted milieu'.

Disrupting the local milieu

The mobility of people like Ira, and to a certain extent also Rolf, impacts on other people's lives. The liberating 'uprooting' of their milieux potentially contributes to the disruption of 'local milieux', milieux that depend on the 'constancy of the surrounding social and material environments of action' (Giddens 1994: 92, 101). Their physical mobility in terms of moving to and from places in search for the appropriate social environment, but subsequently also their symbolic mobility expressed in their rather reserved or relaxed (non) engagement in localized social relations, contributes to the ephemerality and transience of life in metropolitan neighbourhoods.

People in London are well aware of this ephemerality and transience in their local surroundings. In particular, the older generations, who still remember London's close-knit street village communities before the Second World War, experience this transience as 'disruption of local community' and 'decline of community spirit'. Herbert, the retired post office branch manager, who was born and bred in the Elephant and Castle area of London, describes life in one of these street village communities as follows:

> In these working class areas, every street or every two streets were like a village. Everybody knew each other within that street, and perhaps one or two of the neighbouring streets, but, unless they were related or you met

them in employment or in some other way, you didn't necessarily know the people of the surrounding streets so well. So it were like small villages, you lived together, you knew each other's business to an extent. And if anyone died, someone would go around for a collection of flowers, and the blinds would be drawn when there was a funeral. If there was a celebration like the, the first one that I was aware of was the Silver Jubilee of the King and Queen in 1937, a street party would be organized and you'd have a bonfire, and there were all the tables lined up in the street, and there would be fancy dress parades etc. . . . But they would be, each street or each small cluster of streets would be very much like a village.

One can still sense Herbert's nostalgia in this description of a largely vanished street village community, which provided him with a local milieu of concentric circles of familiarity, based on the continuity of generations and shared customs. Accordingly, he is rather resentful about seeing these communities being eroded by people with rather mobile life styles, and subsequently less commitment to local affairs. He describes the transition of London neighbourhoods during his life time as follows.

And I think one of the things that have happened to London now, and I'm jumping a bit here, is that's no longer like that. Because when I was young, people lived in these streets for generations, one generation was succeeded by another generation. But now the population in London is so transient, that they move in and they move out again, you know, there is not the same association with a particular area and the same road anymore.

Herbert identifies the 'Blitz' during the Second World War, when whole neighbourhoods were bombed out, as the beginning of the end of London's street village communities. Subsequent attempts to resettle street communities in the suburbs could not restore the community spirit of 'those days'. Herbert's family was forced to move to the South London suburb of Morden, allocated to a 'big anonymous estate' in which 'there wasn't this sort of get-together spirit . . . the help'. His South London childhood and youth thus reflects what Young and Willmott (1962) have outlined in detail as the large-scale re-housing process of London's East End working class communities. However, Herbert managed to temporarily regain some of that street village community life when he and his bride moved to his mother-in-law's place in North London.

Where we lived in Stamford Hill, there were all white families down the street. And they were all like a street village, everybody knew everybody else, there were successive generations grown up, they knew each other, they knew each other's children, they knew each other's parents, they played in the street together . . .

Yet, the influx of migrant labour from the Caribbean and Asia, hired to run and maintain London's public services (Merriman 1993: 7f.), was to affect Herbert's neighbourhood in Stamford Hill too. The former British Empire imploded. What King (1991: 141), as a consequence of that 'implosion', describes as 'all of Greater London's thirty-two boroughs becoming more cosmopolitan', is experienced by Herbert as disruption of the local milieu he just described in such idyllic colours. His resentment is not simply aimed at the presence of black and Asian neighbours, but directed towards the, as he sees it, rather deviant attitude of these newcomers, disobeying the established norms of the 'indigenous population' of Stamford Hill.

> But then coloured people started moving in to our street. Now, there was a certain friction, because coloured people, for one thing, have a different attitude towards life, towards noise and playing music loudly late into the night. And that causes, that does cause conflict. And so therefore people started to move out, not necessarily because they object the fact of black colour, but, but it doesn't have the same discipline, say as you have.

Herbert was amongst the ones who moved. Once again he attempted to regain a local milieu by moving out of London, to one of the villages just outside London's 'green belt'. This village, as far as Herbert is concerned, has sustained the community spirit that characterized the London of his childhood.

> But coming to this village I regained a bit of what it was like in London in those days, when London was a marvellous place to live.

With its one pub, one school, one shop, and the same neighbours over the last 29 years, Herbert has found the almost ideal typical local milieu, based on the stability of local social relations rooted in the continuity of the local setting and the adherence of everyone to shared values. Accordingly, he feels settled in his milieu. Herbert and his wife 'got involved with things in the village'. He is a member of the parish council, and is a long-standing member of the Parent–Teacher Organization at his youngest daughter's school. Moreover, Herbert can sense the generational continuity of his life-world via the work of his wife.

> My wife, for the past 20 years she has run the village play group . . . So most of the children up to the age of 25 now, she has had passed through her hands in the play group.

It gives Herbert an additional feeling of being 'at home' when these young people acknowledge him and his wife when they go for a stroll in the village, or when he goes for a drink in the village pub. It is not surprisingly then, to see him watch rather cautiously over what he calls, with some irony, his 'last retreat'. For, there are first indications that this village community, that provided him with a fairly sheltered social life in the immediate vicinity of the world city, is going to be challenged by the same processes that once made life in the metropolis unacceptable for Herbert.

> Recently, the big issue in the village has been a planning application by an Asian religious organization . . . to built what they call a retreat, on a large area of wooded and open farm land to the south of the village.

Herbert has opposed this request as a parish councillor, written to local authorities, and campaigned in the village against this 'temple'. However, a quiet sense of defeat comes through in his account of this planning request. He knows there are changes going on in the 'outside world' that will in the long run affect his rural shelter.

> My neighbour's son, who doesn't live here, he lives in the city, said to his mother about the temple coming into the rural world, 'You are now coming into the true coloured world we live in, because we have to put up with strangers in our midst all the time, you know, and you have not'. Which is a point, though. To a certain extent we are isolated, we're cut off. But, of course, we read about it.

Herbert's life trajectory, as we have accounted for it so far, is characterized by successive attempts to maintain or regain a local milieu in the increasingly cosmo-politan and transient social environment of London. Compared to Ira's and Rolf's cases of 'progressive uprooting', Herbert's narrative suggests something like a 'conservative rebalancing' of milieu and environment. His case clearly reveals that physical mobility does not necessarily go hand in hand with 'uprooting' in terms of gaining a certain open-mindedness through leaving behind the moral constraints of a given practical milieu.

Mobility in Herbert's case is not one that aims at 'uprooting' in the previously discussed meaning of the term. It is better described as the forced mobility of a disrupted local milieu, driven by the attempt to re-establish a lost status quo. Thus, the considerable dynamic between milieu and social environment in this case is of essentially nostalgic character. It is stirred on by the repeated disruption of Herbert's practical milieu, caused by forces beyond his control, and leading to subsequent attempts of establishing yet another practical milieu that carries a milieu structure that reflect values of local community. Thus, although Herbert was forced to relocate on more occasions than, for instance Rolf, the term 'uprooting' does not really apply here.

Unlike Ira and Rolf, Herbert is unable to see his own milieu in a wider perspective. In other words, unable to transcend his own local milieu towards a changing cosmopolitan environment, he fails to gain 'world-openness' in Scheler's sense. He does not just object to the physical presence of the 'other' in his milieu, but clearly senses the threat that the 'others' poses towards the long-established 'moral environment' of the 'indigenous population'. This shows, for example, when he complains about their deviant attitude regarding 'noise' and 'discipline'. Herbert lives in the shelter of his own narrow-mindedness, maintained through a succession of local milieux in which there is no place for people other than himself. The 'others' are only allowed to enter into his little world if they are prepared to fit into the rigid values structure that directs his milieu.

> I think, that if the numbers are kept within reason, and the people live according to your rules, speak your language, and work honestly, you know, work in a job, then I feel you can get on with them. Yet when they don't live within your rules, when they speak their own language and won't speak yours, and you can see that they are not interested in working for a living and are living on social security, then you get social unrest, then of course there is crime and violence there as well.

In comparison to Rolf, and Ira to an even greater extent, Herbert does not see the increasing microglobalization of London's life-world as an option to 'uproot' his milieu, but instead considers it as a constant threat to a milieu rooted in localized social relations and morals. As this becomes increasingly difficult to maintain in a context of transience and internationalization, his local milieu is in a structural conflict with the microglobalized life-world of London, which by nature spells local transformation and disruption of localized milieux. Consequently, as Herbert has remained throughout his life a 'local at heart',[1] he was bound to leave London sooner or later to resolve this conflict, even if only on borrowed time.

Finally, all three milieux investigated in this chapter contribute to a clearer view on the 'homeless mind', an issue raised by Berger et al. (1974). Both Ira's and Rolf's narratives suggest that 'uprooting' a milieu from a given social environment does not necessarily imply a loss of meaning or 'metaphysical homelessness'. It seems the opposite applies here. Rolf now feels more 'at home' in London and with himself than ever before in his life. The symbolic situatedness he found in the shared 'moral environment' of London's gay community has given him also a new perspective on the metropolis. Ira, in turn, left her milieu on the council estate precisely for the option to engage in a 'metaphysical odyssey' to find out 'who I'm really like', in other words, to follow her own values rather than the expectations of parents and estate community. In this sense then, each of the shared houses she lives in during her journey are more 'home' to her than the estate she left behind for the simple reason that they allow her 'to be a bit of an individual'. These illustrations clearly run contrary to the assumption of Berger et al. that for the uprooted individual 'no succeeding milieu succeeds in becoming truly home' (1974: 65).

'Uprooting' as it surfaces in Ira's and Rolf's London narratives implies liberation, realization, choice. The loss that is implied in leaving a familiar environment based on locality and family is compensated by familiarity based on choice and association with the like-minded. However, it is in Herbert's narrative that we can realize how closely ontological security and self-identity (Giddens 1993) can be linked to the constancy of a local social and material environment. Herbert is the only truly 'homeless' character, insofar as his localized milieu stands in structural conflict with the microglobalized life-world of London. Therefore, in the long run, all his attempts to realize his milieu, with its emphasis on localized relations and 'indigenous' norms and values, is bound for continuous disruption. In this sense then, Herbert lives, what we, following Scheler (1957: 355), could call a 'tragic milieu'.

All three cases however, indicate something which shall be pursued in the following chapters. Both the 'uprooting' and the 'disruption' of milieux illustrate a far-reaching overall process taking place in the context of global mobility and microglobalization respectively: the delinking of 'locale' and 'milieu'.

5

THE DELINKING OF LOCALE
AND MILIEU

It is a poetic illusion to assume that the world is shrinking because communication improves. In reality the world of each of us constantly expands because, as we carry on, we find it necessary to deal with more and more people, in more and more places, with a greater number and diversity of problems.

Jean Gottmann (1989: 66)

Gottmann's statement aptly captures one crucial ambiguity inherent in current processes of globalization. While the world technically becomes a smaller place, with new means of transport and communication increasingly incorporating people into a single global society, this process at the same time implies increasingly extended, complex, accelerated and transitory everyday lives for most people. What can be described as 'time–space compression' and 'collapse of spatial barriers' (Harvey 1993: 284, 293) on the one hand, implies, on the other hand, that people find themselves thrown into 'accelerated and ever wider-ranging trajectories' within the global cultural economy, be it as labour migrants, expatriates, tourists or members of different other transnational cultures (Lash and Urry 1994: 31). In a similar vein, Giddens argues that disembedding processes in late modernity lead to comparatively 'regular and dense forms of mobility'. As a consequence, the 'primacy of place' as the focus of people's lives and their social relations becomes increasingly undermined (1994: 102f., 108).

The previous chapter, which analysed the 'uprooting' and the 'disruption' of milieux, already pointed towards the 'delinking' of locale and milieu as a crucial feature of disembedding processes. This chapter aims at a further investigation of this tendency in the context of the mobilization and, subsequently, extension of milieux. What becomes clearer when looking at features of the mobile and the extended milieu is the difference between locale and milieu as distinct forms of situatedness. For matters of clarity, locale, following Giddens' (1994: 18) definition, 'refers to the physical setting of social activity as situated geographically'. Milieu,

60

on the other hand, was defined earlier as the relative stable configuration of action and meaning in which the individual actively generates and maintains a distinctive degree of familiarity and practical competence. The argument of this chapter then suggests that disembedding processes in their consequence imply a 'Copernican Turn' from locale-based forms of situatedness towards milieu-related forms of situatedness.

Mobile and generalized milieux

Global mobility is certainly an uneven phenomenon, affecting some milieux more than others. However, there is something to Lash and Urry's (1994: 29) assumption, that 'the professional–managerial classes of the advanced societies are the most footloose'. Following the global flows of capital and finance, their everyday lives are connected with a multiplicity of places, on a more or less temporary or even transient basis. Paraphrasing Park we could ask, Are these the new 'Hobos' of global society, blessed with 'untremmelled mobility', but deprived of 'home' (Park 1974: 156ff.)? From this perspective it makes sense to have a closer look at the milieu of someone belonging to the managerial jet set, in order to find out more about the further implications of the delinking of 'locale' and 'milieu'.

Harold, the 'global business man', works as what he calls a 'firefighter' for the European branch of a New York-based transnational company. Formally based with his 'pigeonhole office' in London, during the week he travels all over Europe and occasionally, if the 'reporting structure' requires, back to New York. He says his job 'has always been travelling'. Travelling is not only a daily feature of Harold's job, but follows also a rather arbitrary pattern, insofar as he follows up 'accidents', both in the literal and metaphorical sense, in the company branches scattered all over Europe.

> A lot of my job is 'firefighting', that is urgent matters that need to be taken care of relatively quickly. For instance, in a plant somewhere a process line stops working, OK. I don't have three weeks to make an appointment, I go there next day, *no matter where I am*, (my emphasis) I change my travel schedule and go next day to that sort of fire, OK . . . Another thing would be that we have decided to make an acquisition. I may have to look at that company, and a quick report goes to New York. I return to London, and suddenly there is a reply, and I have to go to one of the other companies that is associated with the acquisition, and we have to put together a small project, a project of some kind to see how these two fit together. So this is called 'firefighting', OK.

There is a strong indication of a radical break up of the local situatedness of daily routines in Harold's narrative. There is no single place around which his weekly activities are organized. While at least returning to his London base on most weekends, during the week Harold is on the move virtually all the time, without having a fixed base from which to start and to which to return. What Harold's milieu experiences is a shift from primacy of place to primacy of mobility.

Mobility is usually associated with bridging the distance between significant places around which someone's practical relevances and routines are focused. In this regard we could speak of pure or *emptied out mobility*, as it means the relocation of the individual in space, during which the individual's milieu is more or less frozen until the destination is reached and the individual can start reactivating and reorganizing relevant activities around the locale at hand. This feeling of 'doing time' with your hands tied until arrival at the destination, is something everyone who has spent time sitting in an airport lounge can relate to.

In Harold's case, however, there is a shift towards *mobile situatedness* and content-full mobility. Not only does he spend most of his time somewhere between places, but travelling between places for Harold is actually a period of rest, of restructuring relevances and actually getting on with things. With the lack of a proper base, the time spent in the air, paradoxically as it may seem, becomes the focus of his milieu. The plane is where he is actually relatively safe from having his activities redirected to somewhere else. This is where he does crucial paper work but also gets some sleep, things which would normally be associated with locales like the office or home. He is quite aware of the strangeness of this arrangement.

> It's the kind of job, OK, maybe if it was occasional travel, that would be different, but due to the sort of constant travel I have to work during the travel to achieve, you know, my goals. That means a lot of times when I go somewhere I take my paper work with me on a flight to Munich, for example . . . Maybe if it was occasional travel I wouldn't take any paper work with me, because I knew in two days I'm going back to my office and I can do it there, but I don't have that, and that's the problem.

When, as in Harold's case, mobility is not an occasional but rather permanent feature of someone's milieu, a particular challenge is posed to the individual's ability to generate situatedness and familiarity without having the constancy of a locale and its social relations to rely on. Instead it would appear that the skill to generate 'back regions' becomes crucial. Goffman (1990: 114f., 122f.) refers to 'back regions' as a semi-private sphere where the individual can relax from the demands of public performance, but also attend to personal needs such as sleep. While 'back regions' are normally marked off from 'front regions' in a material way, or even embedded in the privacy of the individual's dwelling place, this comfort, as we can detect from Harold's narrative, cannot always be assumed. Where there is no obvious material setting, the individual has then to rely on the ability to 'symbolically cut it [a region for relaxation and regathering] off from the rest of the region'. According to Goffman (ibid.: 130), 'by invoking a backstage style, individuals can transform any region into a backstage'.

As mentioned earlier, Harold is forced to use the plane for serious work as well as sleep. This seems to be a rather extreme form of transforming public space into a back region, given that the personal space he can justifiably claim is fairly restricted, and the contact with other people is rather immediate (Goffman 1972: 52ff.). Harold himself is quite aware of the awkwardness of those situations. But in order to preserve some privacy and control on his daily travels around Europe, he is prepared to stick out as a somewhat 'regressive character'.

> When I'm travelling I don't associate with the rest of the people. I don't know, I don't try to make friends, I don't try to make conversation. Maybe I isolate myself a little bit, but I just don't associate . . . It's the kind of job I guess. Maybe if it was occasional travel, that would be different, but this sort of constant travel . . .

Thus, we find that under conditions of rather permanent mobility, the milieu can find situatedness by reverting into some kind of 'makeshift situatedness' provided by the individual's more or less successful backstage arrangements.

But not even a jet set man like Harold is travelling all the time. Now and then he touches down in locales other than the airplane. Yet, these locales and localities in which he finds himself during the week are not of his own choice and in no predictable sequential order. Instead, almost every day Harold sees himself attempting to realize his milieu in yet another place. In this regard Harold's mobile milieu clearly indicates a delinking of local and milieu, and, consequently, the primacy of milieu over locale. It is the same set of practical preferences and interests that Harold attempts to realize in alternating local environments, thereby, at least to a certain extent, remaining in control of his daily conduct. Harold calls this set of preferences and related routines, acquired over the years, his 'travel sense'. It includes, amongst other things, a feel for convenient travel arrangements, but also what he calls a 'package sense', referring to the anticipation of clothing requirements while away from his base in London, and also a 'sense for cities', reflecting the convenience and safety standards of the airports, the hotels he tends to stay in, and the potential these localities offer for realizing his habit to take a walk after dinner, before going to bed. Each of those 'senses' consists of routines that reflect Harold's practical preferences and interests, and a related stock of practical knowledge, waiting to be activated for realizing those relevances in practice. Harold gives some insight into his 'travel sense' as follows:

> Travel is a trick, OK. Somehow you develop a sense for it. You tend to travel a lot at night, because normally at night it's easy to catch a little bit

of sleep on the airplanes. If you travel late, you tend to get into the airports late, and there are generally more taxis waiting, you don't have a lot of people. I'm a late traveller, OK. I can do a bit of work, I can do a little sleep, and it's not such a hassle. Also, when I arrive somewhere, I don't want to waste time at the airport finding cash machines, OK, so you tend to carry with you some $25 of various currencies for taxi fees.

Harold's mobile milieu illustrates Scheler's rather important differentiation between 'structure of the milieu' and 'practical milieu' (1973: 139ff.). The milieu in its structural make-up of relatively stable preferences and interests, literally travels with the individual. What Harold calls his 'travel sense' is an expression of this milieu structure. To the extent to which these relatively stable 'relevances', to use Schütz's complementary term here, can be realized in the different places in which Harold finds himself, they will be constitutive structures of his alternating practical milieux. This can be illustrated with regard to what Harold calls his 'walking pattern'. Aware of the health risks of his life in the fast lane, Harold insists on taking a lengthy walk after dinner, wherever he might be located that day. Partly through relying on experience, partly through checking with local people, he has developed a 'pattern' applicable to potentially any city.

> Where can I walk, is a question you ask a lot of times. I walk a lot at night, no matter where I am. Normally after dinner I like to take a walk, OK. And I always check, where can I walk, is it safe to walk here, is it safe to walk there? So I tended to develop like little walking patterns . . .

In analytical terms, Harold's mobile milieu can be described as *disembedded relevances of action* (Schütz 1967: 283f.). That is, the set of his practical interests is, in the first instance, not directed towards a particular local environment; nor is it stabilized by being interwoven with the patterns of social (inter)action evolving around a particular locale. Instead, the relevances of Harold's milieu, like his 'walking patterns' can be adapted to potentially any place.

One crucial aspect of this disembedding of practical relevances is the acquisition of an also disembedded 'stock of knowledge', meaning here the 'sedimentation of all our experience of former definitions of previous situations' insofar as they are relevant for dealing practically with the situation at hand (Schütz 1966: 123). In Harold's case this stock of knowledge will, to a small extent, reflect the particularities and specific features of a local setting. It will contain modifiable configurations of typically relevant access points for conducting specific types of action in different environments. In technical terms, it will comprise of patterns of the 'typically alike' rather than 'knowledge of acquaintance' (ibid.) With regard to the example of taking a walk, in none of the many walking routes that he has taken over the years in different cities will Harold develop the same type of familiarity that would allow the lady who walks her dog each night in the same spot of the same park, to go virtually blindfolded. On the other hand, familiar with the general patterns of the urban landscape of most European cities, Harold will be able to repeat his walking routine relatively easy without having to go through the process of 'getting familiar' with the sophistications of any new park he encounters. His relevant knowledge is transferable between places, as it refers to typical features of a routine activity rather than the peculiarities of a local setting. Harold's 'walking pattern' is thus possibly best described by Hannerz's term 'decontextualized knowledge', as it 'can be quickly and shiftingly recontextualized in a series of different settings' (1992: 246).

So, there are certain constitutive routines of his mobile milieu, like taking a walk after dinner, for which Harold depends on his ability to deploy his considerable 'decontextualized knowledge' to realize these routines in a more or less appropriate setting. However, with regard to other practices he can rely on what could be described as *generalized milieux*. By this I mean prearranged local settings that provide or serve the basic needs of the mobile individual in a standardized manner. Generalized milieux are based on the anticipation of the structure of relevances concerning general daily routines like eating, sleeping and transportation, which subsequently finds expression in certain standardized material settings such as hotels, airports, and fast-food outlets.

Moving through this circuit of generalized milieux gives the mobile milieu a form of situatedness other than that evolving around a fixed locale or specific locality. While, on the one hand, any of these standardized settings more or less fulfils the basic needs of the global traveller, there is, on the other hand, scope for individual preferences, and even a certain sense of familiarity. Harold, for instance, mentions his 'airport sense', based on his judgement of efficiency and safety standards. In this regard he prefers, for example, Frankfurt to Brussels. Similarly important to him is the 'hotel sense' he has developed, with preferences depending on whether they accept his credit cards, serve light food like salads, and are conveniently located near some kind of park. Over the years Harold has thus found 'his' hotel in almost each major European city.

> It's also a hotel sense you develop, OK, you really develop a hotel sense of places you prefer to stay at. All over this period of years I developed this strong attachment to the hotels I stay in. I like to stay, I absolutely refuse to change, you know. Most people I travel with don't stay in the hotels I stay in, but I refuse to change.

What is rather obvious here is the link between permanent and unpredictable mobility on the one hand, and the compensating – because stabilizing – role of generalized milieux on the other. As Harold literally can never be sure where in

Europe, or even America, he will spend the next night, he tries to compensate for this imposed unpredictability in his milieu by a conscious decision for consistency as far as it is in his control. Staying at a convenient hotel of his choice, despite it being not more than a stop over, gives back some feeling of situatedness. Furthermore, generalized milieux seem to leave scope for individual preferences, as Harold and his colleagues have obviously different opinions as to what is the best hotel in town. This observation goes somewhat against the dominant view of these standardized settings as 'dehumanized' (Ritzer 1993: 17) or 'non-places' to which individuals are connected in a uniform manner (Augé 1995).

That these generalized milieux, despite their standardized setting and social anonymity, have an implicit potential for the maintenance of the individual's milieu, also comes to the fore when Harold talks about his eating habits, as already indicated. To maintain his rather healthy attitude towards late night dinners turns out to be a major problem when travelling through Europe's corporate culture, where participation in business lunches and dinners is good form. For Harold the best way to escape these culinary pressures of his respective host environments is to escape to McDonald's, whenever the corporate etiquette allows.

> Maybe unbeknown to them, the last three nights I had to eat with somebody else, OK, and the meals have been big and long, then sometimes I just prefer to go into a McDonald's, have a nice fish burger and salad, and a milk shake, and that's it.

As paradoxical as it might appear, the above sequences indicate that generalized milieux can offer crucial assistance to the mobile individual in the attempt to maintain a personal milieu while negotiating a way through the embedded routines of local culture(s), whether culinary or otherwise. The appeal of these generalized milieux seems to lay in the fact that they offer a compromise between 'intrinsic' and 'imposed relevances'.

Following Schütz's analysis of 'routine activities' (1970: 144), it could be argued that generalized settings such as hotels or fast-food outlets assist the realization of *decontextualized relevances of action*. This means that relevant issues concerning general daily activities such as eating, hygiene, resting, and so on, have been isolated from their wider motivational, topical and interpretational context. In other words, they have been decontextualized from respectively unique 'biographical situations' and find their materialization in generalized settings like the ones discussed above. This, then, implies that the routine activities realized in these settings are no longer considered problematic in themselves and in the individual's wider horizon of activities. Clearly defined and anticipated in their 'functional character' as 'specific means' for 'specific ends', they do not need further investigation to be integrated as unproblematic elements into the individual's course of action.

66

Two implications follow from this. First, these decontextualized relevances of action are of 'fixed' character. This allows their materialization in a standardized setting, but they also acquire the taken-for-granted character of tools for other, more problematic, activities. In Schütz's terms these 'decontextualized relevances' become subordinated to other 'relevance systems . . . in relation to which they function just as a specific means to bring about specific ends of higher order' (ibid.). To relate back to Harold's narrative, he certainly does not want to face the daily challenge of finding a place to rest and sleep each time he arrives in yet another city. With regard to these tasks, to rely on the standardized convenience of generalized milieux enables him to concentrate on more important issues on the 'firefighter's agenda'. On the other hand, their tool character means that these generalized settings can be potentially incorporated into the milieux of rather different people. Thus we can assume that Harold will have to share the standardized convenience of the McDonalds restaurant with the fellow traveller, the commuter, the student, the shopper, or the local character on the lookout for a free newspaper.

Second, the fixed character of decontextualized relevances actually allows their compatibility with the mobile milieu. Generalized milieux like the hotel or the fast-food outlet guarantee the immediate integration of these fixed, or materialized, relevances of action, into the wider system of relevances carried around with the mobile individual. This is nicely expressed by Harold when he refers to the convenience of 'checking in and out of hotels'. He can rely on the fact that certain fixed relevances will be taken care of in these generalized milieux, ready to be (re)integrated into his milieu whenever and wherever needed, and without him having to do any maintenance work.

However, the suggestion that a mobile milieu, like Harold's, might be more reliant on generalized settings should not obscure the fact that any milieu is, in the end, dependent on the repair and maintenance work of the individual. This perhaps applies to the mobile milieu even more so than to any other milieu. Deprived of its local focus, the consistency and continuity of the mobile milieu is down to the individual's active effort and motivation to keep all relevances literally 'in his grasp' (Schütz 1966: 131). Thus we should perhaps not be surprised to see in Harold's case the reliance on generalized milieux, and thereby in the reliability of others, balanced out by a keen sense to personally 'handle' the practical side of his otherwise fairly unpredictable travel arrangements.

> Most of my travel arrangements I tend to make myself, OK. Some I make through my secretary, mostly longer term things, I let her handle. But I never ever really learned over the last 30 years to trust anybody to handle my travel arrangements.

In sum, what we can detect from Harold's account is that the mobile milieu relies for its situatedness both on the access to generalized milieux and a stock of decontextualized knowledge. Both these 'devices' support the individual's attempt to focus the personal relevances of everyday life without being able to rely on the constancy of a specific locale. The following section will take this argument one step further by suggesting that the uprooting and mobilization of milieux finds a logical continuation in their extension.

The extended milieu – between 'here' and 'there'

So far the argument has been concerned with the uprooting and increased velocity of milieux. The basic assumption until now has been that milieux travel from place to place. Implied in this assumption is the perception of the milieu as some kind of cocoon surrounding the individual. The following argument suggests that the milieu itself can potentially be stretched across infinite time–space distances. The notion of an *extended milieu* partly summarizes, partly goes beyond what has been said so far with regard to the overall process concerning the delinking of locale and milieu.

We can start with the simple observation that many of the milieux under investigation here have more than one life-centre. This seems to be the norm rather than an exception when looking at the biography as a trajectory in time, where different status passages, such as birth, education, work, and retirement, go hand in hand with spatial relocation. However, when looking at the milieu as a spatial configuration we can increasingly note the coordination of life-plans, and in some cases even the organization of daily routines, around a 'multiplicity of significant places' (Waldenfels 1985: 195ff.).

Harold, for instance, lived in the United States, the Middle East and South Africa before he came to live in London. The family kept their house back in Texas throughout that time, and go there at least once a year for a long holiday. Harold's daughter has returned there to study, while coming to visit her parents in London a couple of times each year. After his retirement Harold wants to return to Texas to raise cattle. During the week he is travelling from one European city to another. Even when actually in London, Harold's daily life centres around two other significant places apart from his place of residence, the American School and the Lutheran Church in the City.

In contrast to a localized milieu, Harold's milieu, which revolves around a cluster of 'significant places', can thus be described by a certain *decentredness*. This implies, in the first instance, the loss of a particular locale as the single focus of all life-plans and daily routines. We could say that in this scenario the 'biographical situation' (Schütz 1970: 167f.) loses its place identity. But as we could see in Harold's, admittedly extreme, case of a mobile milieu, this decentredness can also affect the continuity of daily routines and the maintenance of personal belongings. This is illustrated when

68

Harold talks about being forced to 'keep a double set of personal belongings' which takes into account his rather permanent and unpredictable pattern of travel.

> When I go home, my shaving things, new T-shirts, new undershirts, new shirts . . . are folded for me already, so that I only have to exchange them quickly, the old comes out, the new goes in.

What this suggests is that with the decentredness of the individual's milieu goes the multiplication and fragmentation of what, following Goffman (1972: 51ff., 338), could be called the 'symbolic territory of the self', – or, all material and non-material expressions of the self, detachable from the individual's actual *Umwelt*, yet effectively experienced as an essential part of the individual's milieu.

This can be further illustrated by looking at a similar pattern in the milieu of Barbara, who in the third chapter was introduced as the 'expatriate'. Before taking the temporary post in London she moved most of her belongings from Manila to Nashville, where she owns a house. Also in Nashville, she has arranged a safe deposit box, which contains her will and other important documents. A friend acts as trustee for these rather crucial personal belongings. To London, where with all likelihood, she is going to stay for about two years, Barbara brought detachable things that are nonetheless 'meaningful to me and other people'. These include pictures and letters from her children, newspaper clippings from the Philippines, her son's Ph.D. proposal, and a radio that 'gives me a chance to have some control over what I hear, what I experience'. How important these fragments of her symbolic *Umwelt* are for the situatedness of her practical milieu in London can be sensed from the following remark:

> So, it's like my room and these little things, that's my little world right now. Actually, the pictures of my children, I took them out of my purse, because I would rather not lose my identity with my family and friends, money is not so important.

There is one other aspect of decentredness that becomes evident by looking at Harold's and Barbara's narratives. We have to recall that the milieu is not just a configuration of action but also meaning. As such it transcends the individual's immediate 'here and now'. The milieu in this sense is not confined to the individual's *Umwelt* in Goffman's definition (1972: 297ff.). Milieu was earlier referred to as the individual's value-related environment. It extends beyond immediate practical interests and the task at hand into the realms of hope, desire, fear, expectation. The individual's 'biographical situation' thus also generates the milieu as a shifting 'affective field', which is not confined to the milieu as 'perceptual field', nor does it necessarily coincide with milieu as 'field of action' (Waldenfels 1985:

195ff.). This is illustrated by the simple observation that, more often than not, the place of bodily situatedness is not necessarily the focus of affective situatedness. Harold might be located somewhere in a hotel in Europe, but his mind is with his son sitting exams in the American School in London or his daughter having an interview for a place at an American university. We encountered an extreme case of this decentredness of 'field of action' and 'affective field' earlier in Rolf's narrative about his initial time in London. Reluctantly relocating to London, the affective side of his milieu was centred around 'home' in Germany and focused on coping strategies of how to 'survive' three years in a place not of his choice, and accordingly, avoiding any emotional attachment to London.

Consequently, in view of the multiplication of 'significant places', the decentredness of 'field of action' and 'affective field' – and subsequently the simultaneous multiplication and fragmentation of the 'symbolic territory of the self' – it can be argued that milieux inhabit space rather than places. The assertion of milieux inhabiting space has several dimensions. There is, first, the increased mobility of milieux in *territorial or geographical space* due to global means of transport. These open up a 'field of action' of a routine territorial extension unknown to former generations. We saw Harold, for instance, touring the whole transatlantic region without always having a precise destination for the next day, yet nevertheless being able to return 'home' to London most weekends. That this new everydayness of global mobility is not confined to the managerial jet set can be illustrated with reference to Hannerz's (1992: 238) observation of Lagos market women using London-bound planes as a shuttle service for their commercial activities.

Second, however, the extension of milieux into space implies more than mere physical mobility between 'significant places', technically bridging them via 'abstract systems' (Giddens 1994: 80). Milieux have a direction, deriving from people's shifting 'disposition'. From this perspective then, places are no longer entities in a uniform 'empty space' (ibid.: 19), but they become 'significant places', interwoven in the meaningful configuration of biographical relevances.

This link between 'significant places' within the milieu's 'affective field' comes to the fore most clearly where there is no high density physical mobility between them, yet where distant places and people have a crucial impact on the emotional situatedness of a milieu. Ulla, who was introduced as the 'settled cosmopolitan' in chapter 3, described herself as a 'Londoner by birth' with family ties extending towards Sweden and America. While she would refuse to live anywhere else other than London, at the same time she acknowledges the importance of the emotional links to America, in particular. It turns out that without these links she would have had difficulty making it through emotionally difficult times. Especially during a time in the early 1950s when Ulla was left to nurse her ageing parents, her brother and sister provided crucial material but also emotional support by providing the financial resources for a yearly trip to New York and Newfoundland. The vision of these holidays and sibling reunions were the remedy that kept her going in London during that period:

> My brother always sent me money to get out to see him, to have a holiday. And I say, without those holidays I wouldn't have survived at all . . . He either sent me money or a round-trip ticket. I would then go to Newfoundland, stay with my sister perhaps for a week or two weeks, see all my friends there, and then fly down to New York to be with my brother.

What should be recognized here is not so much the fact that Ulla actually travelled to the States, but that the vision of leaving London at least once a year made her keep going in view of the rather distressing circumstances in her immediate environment. There is clearly an impact here of distant places and people on Ulla's hopes and expectations, thereby giving her milieu affective situatedness. How essential the continuity of affective links across distance can be for the stability of a milieu can be revealed by further outlining Ulla's narrative. Far from the contacts ebbing away after her sister's death, there are perhaps even stronger links now between London and Newfoundland, which, despite Ulla's advanced age, encourage regular visits in both directions almost every other year. She has made real friends there, with geographical distance not being an obstacle. With phone calls and 'pigeon post', that is, someone coming over from 'there' and posting a bunch of letters 'here', Ulla is familiar with happenings in the little village in Newfoundland as much as with events in her local road in Streatham.

> And of course, when you have all these friends, that's why I want to go this summer, because I've made so many friends there. I've really got to know these friends, you know, have heard all their stories and everything else. My sister being three years older, I always made younger friends, it weren't hers necessarily.

A particularly strong relationship has formed over the years between Ulla and a woman, Anna, who looked after Ulla in Newfoundland when her sister died. Theirs is more than a friendship of convenience, in that it has features of what Ulla calls a 'transatlantic support service in crucial situations in life'.

> There was the daughter of that couple, Anna, who looked after me when my sister was dying. She put me up in her guest room and really comforted me. Since then we are friends and keep in touch, which is the important thing . . . So one year, for instance, I flew over and looked after their home while they were having a holiday in Ireland, you see, that kind of thing.

71

While in Ulla's case the extension of the affective field of her milieu clearly plays a stabilizing function, the story of Nicos, the 'local entrepreneur', provides a different picture. In his case the strong emotional link to Cyprus has a rather unsettling impact on his milieu. His case is of additional interest as it reveals the impact of 'mediascapes' on the way in which significant places are linked in Nicos's milieu. Appadurai defines 'mediascapes' as 'image centred, narrative based accounts of strips of reality . . . out of which scripts can be formed of imagined lives, their own as well as those of others living in other places' (1992: 292). Nicos is caught between two of those reality-defining images. On the one hand there is the image of the 'golden streets of London', which defines the expectations of his relatives back home about Nicos 'the potential returnee'. Their expectations implicitly determine Nicos's practical disposition towards London, as he is 'doomed' towards financial success. On the other hand, he has developed a holiday image of Cyprus, in contrast with the harsh and rather anonymous life in London. He is caught in a vicious circle insofar as returning home is possible only when he has lived up to images of the wealthy homecomer, while on the other hand he hates London life exactly because of that implicit pressure from Cyprus. Sometimes he would rather throw it all in and stop having his life ruled by an ideology stretching between London and Cyprus, with him caught in emotional no-man's land, happy neither here nor there.

> The ideology is, that you are in London, you should have made a lot of money, you should be somebody, you know. They expect you to go home on holiday with lots of money in your pocket, because they have this illusion about, you know, the 'golden streets of London', you know, this is rich pickings. And, you know, if you spent 10 years in London, or 20 years in London, why haven't you got a lot of money? Are you an idiot? The fact is that you are not an idiot, the fact is that life is hard.

This interview sequence clearly indicates that the extended milieu, as a configuration of meaning across distance, takes in and is influenced by happenings and people in distant places. The individual's symbolic attachment to significant places, due to previous biographical sediments, as much as future life-plans generates an 'affective field' well beyond the actual 'here and now'. Moreover, the extended milieu links distant places and happenings into a symbolic territory that is relatively independent of geographical territoriality, and instead measured by biographical relevance. A quite extreme case of that relative neglect of objective distance was seen in Ulla's account on the 'transatlantic support service', where relatively elderly people would cross the Atlantic in order to engage in neighbourly assistance.

This idea of the milieu being a symbolic territory in which the space between significant places is by no means empty, can be further explored by looking back at

Barbara's life story. We have to recall that there are strong links to the Philippines in Barbara's biography. Before she came to London, Barbara lived for a substantial period of time in the Philippines. And it is her intention to return there after her two years in London. Accordingly, the general disposition of her milieu is geared towards the 'Filipino experience', as she calls it. This disposition supports the continuity of her milieu not just in terms of affective situatedness, but also in terms of the concrete structuring of her practical milieu in London. Her closest friends in London are a Filipino family, and at work she mingles with women who share the 'Filipino experience' in one way or the other. She actively looks out for places and events associated with London's Filipino community. Barbara attends community festivals, and Filipino dishes are a substantial part of her weekly diet. Consequently, it could be argued that Barbara has not relocated 3500 miles between places on two different continents, but moved from Manila to Filipino London – two significant places in immediate neighbourhood in the symbolic territory of her extended milieu.

It goes without saying that the extended milieu is not confined to the links between just two places. As a 'symbolic territory of the self', it can integrate a number of significant places, whereby the significance of these places can alter with changes in the individual's 'biographical situation', in other words, changes in life-plans or immediate tasks at hand. Barbara's life at the moment centres around at least three places. There is Nashville where her children and property are, but also the necessary money flow, and long-terms plans for her retirement. Then there is London, her momentary place of residence, and finally, there is Manila to where she wants to return after her posting in London. No wonder that she feels rather unsure as to which of these localities to consider as her 'home', in the wider meaning of the word.

> Well, I'm thinking of London now increasingly as home, more and more, while I still say 'home' sometimes *there*, but more and more I'm saying this is my home, but I think of home as being more, . . . London than Streatham . . . It *goes all around in fact, goes everywhere*. But there are more things *here* now than there used to be, obviously. Because there are more people I'm with, people I have vast amounts of work to do with. But then again, I get mail or e-mail from home, and I get information from friends and family, and so on, and it makes me think about them. And there are some things that I miss here, and there are certain things that make me feel good to know that they are *there* and I'm *here*, and that's OK, it's not as if I have to be *there* . . . (my emphasis).

What Barbara's narrative indicates is a *poly-centredness* of her milieu that goes beyond the decentredness described above. What is expressed in her genuine confusion about 'here' and 'there' is the lack of a single local focus in her extended milieu. 'Field of action', 'affective field', and technical links overlap in many ways without being focused in a single place. Different life situations and practical tasks will push different places to the fore. Money and insurance problems will be dealt with via a friend in Nashville, the conduct of her daily life will make her more familiar with London, while at the same time Manila is the focus of past good times in her life, and her plans for the immediate future. Yet this is not a fixed distribution of relevances or significance, as her son's upcoming doctoral viva, for instance, will shift the affective situatedness of her milieu from Manila to Nashville, and so on.

Also, despite the fact that there is no proper focus of situatedness, there is a sense of direction in Barbara's milieu, giving stability to its affective field, as well as meaning to the tasks at hand. Her aim is to return to the Philippines, and that intention also structures her momentary life in London, largely evolving around Filipino London. With de Certeau (1988: 117), we could say that her milieu has a 'direction of existence'.

It is obvious, then, that the poly-centredness of the extended milieu does not mean indifference between significant places. Recalling Schütz's idea of the bio-graphical situation, the significance of a place will change according to life-plans, practical tasks, future life-plans. The individual's attempt to generate continuity of the biographical project is thus transferred into a spatial dimension. That these plans, tasks and emotional preferences do not always coincide goes without saying. There is room for conflict and the 'tragic milieu', as is partly evident in Nicos's story.

The extended milieu is not just a configuration of places with fixed positions in geographical space, linked by technical means. Instead, and, again, following de Certeau (ibid.), it can be described as a 'polyvalent unity of conflictual programs or contractual proximities', generating a symbolic space that is 'produced by the operations that orient it, situate it, and temporalize it'.

The convergence of 'here' and 'there'

The different dimensions of the delinking of locale and milieu discussed in the previous sections – uprooting, mobilization, transcendence, decentredness, and finally poly-centredness – can be summarized in the notion of an extended milieu that no longer finds its situatedness in the continuous attachment to a particular locale. The extended milieu transcends the surroundings of specific locales in its 'field of action' but also, and even more so, in its 'affective field'. Moreover, the extended milieu inhabits space rather than place in that it integrates distant places, people and happenings in a meaningful configuration of biographical relevances. Conversely, these distant places, people and happenings can have immediate impact on the individual's actual 'here and now' as well as the general affective disposition of the milieu. This means that the extended milieu evolves around a multiplicity of significant places, whereby none of these places necessarily provides the ultimate focus.

However, the point is that the terms used to describe the extended milieu have themselves become ambiguous in their meaning. 'Here' and 'there' do not equal 'near' and 'far' in the meaning of the familiar and the unfamiliar. 'Distant' becomes problematic in its meaning because 'significant places' are not arranged in concentric circles of diminishing relevance. Significance of places, people and events is not measured by geographical distance. The mobile and extended milieu has to handle the paradoxes of distant familiar places and 'others', as well as unfamiliar immediate surroundings and 'others'.

It can be suggested, then, that what Harvey (1993: 240) describes as 'time–space compression', has its life-world equivalent in the *potential convergence of 'here' and 'there'* in people's milieux. And this is not just a passive consequence of the 'collapse of spatial barriers' and 'imploding spatialities' (ibid.: 299, 304). Instead, it can be suggested that it is people in their milieux who actively generate a unique spatial order that reflects their access to a global life-world. Following Harvey, it can be argued that despite the radical change in the objective qualities of space and time in the context of time–space compression, 'time and space have [not] disappeared as meaningful dimensions to human thought and action'. What has radically changed indeed, is 'how we represent the world to ourselves' (ibid.: 240, 299).

The collapse of geographical distances does not mean the absence of any spatiality from the individual's milieu. Instead, geographical measures are replaced by shifting biographical relevances. The collapse of a fixed geographical order of places is replaced by an 'original spatial structure' deriving from the creative manipulation of 'seemingly contingent geographical circumstances' (Harvey 1993: 294; see also de Certeau 1988: 109). The individual's competence to successfully focus and balance intermingling local, regional and global relevances of action, generates a concrete biographical spatial order in which an imploding, and yet at the same time extending, world makes sense.

This creative manipulation of 'seemingly contingent spaces' applies potentially to both sides of 'time–space compression' as experienced in people's biographies. That is, the extension of milieu towards a potentially global field of action on the one hand, and the implosion of the world's variety and differences into environments of microglobalization, such as London, on the other. Two illustrations might further clarify this point.

Sarah, the 'mobile cosmopolitan', certainly cherishes the metropolitan juxtaposition of difference that London has to offer. She enjoys the wine bars in the West End as much as a visit to Brixton market on a Saturday morning. Yet, in order to really enjoy the different worlds London has to offer, Sarah refuses to be randomly exposed to them. Consequently, she attempts to stamp her preferences onto her involvement with London life. This involves, among other things, to keep 'two different sets of friends', one consisting of City workers like herself, with the necessary money to enjoy the expensive sides of London life, or indeed going off on sailing weekends in Spain and France. The other set of friends operates more on the basis of like-mindedness. It includes former colleagues from her past as an air stewardess, most earning rather less money than her, but providing good company for the more bohemian aspects of London life. This attempt to stay in control of the different social worlds she participates in includes setting symbolic differences where there are no obvious spatial boundaries between 'here' and 'there', in this case referring to the difference between work and leisure.

> So for me there is an enormous psychological difference where I have my entertainment. I don't stay in the centre of London because if I finish work I want to go and switch off and relax totally. The way I do that is to come home and change my clothes, and go out to somewhere that has nothing to do with work. I hate going out straight after work, which a lot of people tend to do. I hate meeting up with people for drinks straight after work and then going to the cinema or something in the West End because I just don't feel that I have changed gear at all, and that's very important to me. So I feel in a way that, you know, that although we are geographically *quite near* the City here, I feel it's *a world apart* (my emphasis).

A similar kind of balancing of commitments between 'here' and 'there' can be detected in Ulla's case. She has equally strong ties with some people in her residential road in Streatham, and with some people over in Newfoundland. There are commitments with regard to special events here as well as there, which at times makes it difficult to plan the near future between conflicting local and transatlantic relevances.

> In fact, I'm going over there very soon. I'm still waiting to book [the flight] as I want to make my dates with Ruth [her neighbour], because Ross [neighbour's husband] has got his special anniversary, hasn't he, you know, when he retires. I want to know the dates before I say I'm going off to Newfoundland on the so and so.

What these sequences indicate is a calm handling of the spatial contingencies of an increasingly (micro)globalized life-world. They show engagement with a global world rather than a retreat from it. The notion of situatedness of extended milieux as described in this chapter therefore refuses the nostalgia that some phenomenological approaches are rightly accused of (see Robertson 1993: 157). In this version, milieu becomes a possible form of 'the concrete structuration of the world as a whole' which Robertson urges social scientists to focus their attention on in order to understand the cultural dimension of globalization (ibid.: 53).

Moreover, if in the context of modernity the uprooting of milieux, and their increased *mobility from place to place*, implied the 'homeless milieu', to take up Berger et al.'s argument here again, then the lived experience of extended milieux, as shown in this chapter, indicates a crucial transformation in society. The increasing everydayness of access to global means of transport and communications in the widest possible sense, and with it the access to a globalizing life-world, allows milieux to inhabit the *space between places*, to remain in a sphere of permanent transition and convergence between 'here' and 'there'. Situated somewhere in between two or more social worlds, the extended milieu does not necessarily need to feel displacement. In this regard it could then be argued that globalization processes realize the epistemological roots of the concept: mi-lieu, *the between of two places* (Rabinow 1989: 129). Thus the 'homeless milieu' and the 'extended milieu' can be seen as life-world indications of the modern age and the global age respectively.

'Home' as a significant place

The mobile and extended milieu outlined in the previous sections emphasizes Giddens point that in the context of disembedding processes the locale loses its primacy as a focus of people's daily lives (1994: 108). However, the delinking of locale and milieu does not imply the *de-localization* of milieux. Instead, we saw the extended milieu evolving around 'significant places'. There is no contradiction between the assumption of a disembedding of milieux from pre-given locales, and their re-embedding around 'significant places'. As shown in the previous section of this chapter, the 'significance' of these places does not designate their 'locale' – their physical setting as situated geographically – but refers to the relevance of these places within the spatial and symbolic configuration of the individual's milieu. To put it rather pronouncedly, rather than the milieu being determined by the constancy of the locale, it is the locale that finds its practical and symbolic significance in the context of the milieu.

The point I want to make here appears to be quite simple, yet is of deep analytical consequence. As long as people's lives evolve from the constancy of a specific locale and the continuity of the local social relations that go with it, there is hardly scope for the individual's active efforts with regard to maintaining a milieu to surface. To speak in conceptual terms, 'locale' and 'milieu' converge into a 'local milieu' (Giddens 1994: 101). It could be argued that the disruption of localized milieux in the context of globalization processes furthered the realization that in an increasingly transient world people will have to actively generate 'their' milieux instead of relying on the constancy of the surrounding local and the attached local relationships. To further illustrate this point, the following section will look at the link between 'home' as a significant place and milieu.

Significant places imply, to a certain extent at least, an element of choice and selection. Mobility and extension of people's milieux imply that more than one place can qualify for the notion of home. No longer does the place of birth automatically qualify for this special symbolic attachment of the individual. While certainly everyone in this transient world will be able to identify their 'home' of some kind, what we imply by this, however, is a plethora of things. Less and less it seems, is 'home' associated with a life-long attachment to a certain locale or locality.

In Sarah, the 'mobile cosmopolitan', we have a vivid example of the rather loose links between 'locale' and 'home'. In her early thirties, working as a head hunter in the City, and formerly employed as a stewardess with British Airways, Sarah has moved in and out of London, depending on the requirements of her jobs. Her current place of residence is in Peckham, South London, which she chose because of 'easy access to the City', 'fresh air', and 'lots of open space'. Her place of residence is first and foremost a 'base' for activities across London and the world, but hardly a place of social engagement. Her friends live 'scattered all over London', so that her social life takes place in 'neutral territory', as she calls the places that are within convenient reach for both her and her friends. Sarah's singing hobby requires a lot of travelling and she likes travelling anyway. So, London in general, but in particular Peckham, with relatively easy access to Gatwick Airport, is a base for travel, rather than to settle into a local community.

> If you live in London, it's very easy to think about going somewhere else, whether it's outside London in the UK, or outside the UK altogether. I mean, this year I've got an awful lot of travel organized already, and I've got weekends in Barcelona, weekends sailing along the coast of France, possibly, trips of course to Boston, and all sorts of things, which are easy for me to do because I'm based here, and I can just sort of buy a reasonably cheap ticket from London and get there easily after work, and just go away for a long weekend.

A similar perspective on 'home' holds true for Ira, despite having her mobility confined to the London area. Her parent's flat on the council estate certainly did not feel like true 'home' to her, so she left. On her journey towards 'finding out what I'm really like', we recall, Ira went for places of residence that allowed her to open up her 'avenues' to people, and taught her about metropolitan life. She certainly did not want to find 'home' in terms of local settledness, but find home to herself, so to speak. In this regard, places were losing their appeal when they became too 'homely'. This does not mean that these places were all the same to her. There are *places of* particular *biographical significance* for Ira, places 'where I became a bit of an individual' as she says.

I think some places, you know, they have a significance for you if they kind of represent a changing period in your life, you know, or a discovery period in your life.

Without needing to know what exactly both Sarah and Ira mean by 'home', it becomes clear from their narratives that 'home' for them derives from the ability to generate a significant relationship to a place, rather than through fitting into a local setting and local tradition. What makes a locality or locale a 'home', depends on how it fits into the changing requirements of their biographical situations.

Ira, now in her late twenties, has moved house at least six times since she left the estate in Battersea. Having lived in areas all over London, she has always managed to find something 'unique' about these areas that allowed her to feel at home. However, she likes these places not because of their local setting or local history; what is important to her is that she is able to relate to the people who co-inhabit this place, and the overall atmosphere that is attached to a locality. She tries to account for these quite diffuse feelings of attachment referring to the uniqueness of Brixton, her current place of residence, as follows:

> I've never come out of the tube station actually at any other place, where I thought to myself, my god, I can't believe these people, you know, I can't believe it. You know, I've come back from somewhere, tired as hell on a Saturday afternoon, and as soon as I'm at the bottom of the escalator I can hear Brixton, I can *feel* Brixton [my emphasis]. Then you are going up the stairs, and there are lots of people on the stairs, and you're putting you ticket through, and suddenly there are four or five people, and they're either asking you for money, or are telling you something about the world or about religion, or they're trying to chat you up, or they're trying to sell you something . . . there is just so many. And then you get to the top, and it's even more of that, you know. I think that's amazing, that's why I essentially like Brixton.

Speaking in more conceptual terms, Ira feels attachment to these place most of all due to the 'affective field' (Waldenfels 1985) that radiates from these places, and the extent to which this is in tune with her own 'vibes', as she says. 'Home' in her case is not primarily related to the familiarity of her own home or her neighbourhood. Ira's relationship to Brixton suggests an extension of the notion of 'home' towards a diffuse affective configuration without clear time–space allocations.

A delinking of 'home' and a geographic notion of locale can also be detected in Sarah's account on what 'feeling at home' refers to, though in her case we seem to encounter a contraction rather than extension of the notion of 'home'. Pressed on this issue of what 'home' means to a much travelled person with a fairly mobile life style, she avoids calling a locality like London or even Peckham her 'home', and prefers instead to give it a personal index, fitting the mobility of her milieu:

> But isn't it [home] related to this house largely, I suppose? I would have the same feelings whether or not it is London I'm coming back to. I mean, for example, when I was travelling a lot with BA, and I lived out in Oxfordshire, then that was coming home to me, but it wasn't London. So, for me its a more personal thing, is more centred on my house, and I feel my belongings are here and, you know, or my independence, if you like, is here. It's *not sort of geographical* [my emphasis].

Looking at both Ira's and Sarah's idea of 'home', it does not seem to contradict mobility both in a geographical and metaphysical sense. For both of them it derives primarily from the ability *to make* themselves feel at home at different places. In both Ira's and Sarah's case 'home' equals 'independence', and neither sees a contradiction in that. A similar association of 'home' with independence and 'freedom' can be seen in Rolf's earlier statement referring to the first flat after his 'coming out' as being his first 'real home'. Also, Barbara's way of making herself at home for the next two years in a rented room in a shared house illustrates the link between mobility and symbolic anchors of situatedness. As we have seen, amongst the things she brought with her from Manila were pictures of her children, letters, books, and her son's Ph.D. proposal. These things are not just personal belongings in a technical sense, but are significant objects, 'things that are meaningful for me and other people' as Barbara says. These things connect her with distant places and people, and interlink past and future events in her life.

There is good reason then, to conclude that the delinking of locale and milieu does not imply homelessness in the metaphysical sense. Instead, we can argue that the mobilization and extension of people's milieux actually brings to the fore the individual's 'very ability' to actively 'identify home' in an ongoing 'construction and organization of interlaced categories of space and time' (Robertson 1995: 35).

As already discussed in detail, this implies, first, the ability to generate and maintain 'home' in the sense of *Heimat* in a biographical succession of places – place of birth, education, residence, work, retirement and other status passages. But more importantly, it seems that this ability to generate home in that wider sense will increasingly require skills to organize one's biography around and between a multiplicity of 'significant places', in all of which the individual is forced to be 'at home' to a greater or lesser extent (Waldenfels 1985: 188).

This process of generating situatedness is also evident when looking at 'home' in the narrower meaning of the place of residence. In this respect we noticed that the individual can generate feelings of being at home on the basis of the affective disposition of the milieu feeling in tune with the 'affective field' emanating from a place. Ira, for example claimed that she could 'feel' Brixton when coming out of the underground station. Moreover, we saw that 'home' can have the personal index of property, crucial belongings, and seemingly trifling yet personally meaningful things that link the individual's 'here and now' with distant places and people.

Taking up Goffman's ideas at this point, we could say that the individual's home extends well beyond the fixed territory of the self defined by fixed equipment and geographical location. It extends towards the 'possessional territory of the self' which includes all 'claims' towards detachable personal things. In a wider sense, this includes distant possessions as well as significant others, non-present, but essentially linked to the individual's 'self', and therefore 'effectively part of his [or her] *Umwelt*' (Goffman 1972: 51, 62, 338).

In this sense, for example, her son's Ph.D. proposal is a crucial part of Barbara's milieu, embodying the link with her son. Invisible as the significance of this proposal might be to others, it helps Barbara to make a little room in London's Streatham her 'home' for the time being. Following Goffman, it seems plausible to call these symbolizations or embodiments of the individual's links between here and there, present, past and future, the *symbolic territory of the self*. The individual's milieu finds its 'home' in this extended configuration of real things and the intersubjective meanings attached to them.

Looking at the situatedness of milieux from the perspective outlined in this section, one can finally only agree with Robertson who argues, following Balibar, that 'in the present situation of global complexity, the idea of home has to be divorced analytically from the idea of locality' (Robertson 1995: 39).

Environments of like-mindedness – St Anne and St Agnes Church

So far in this chapter I have been mainly concerned with the different dimensions of the delinking of milieu and locale. I have argued that the mobilization and extension of milieux does not necessarily lead to the restlessness of 'homeless minds', despite the diminishing embeddedness in the constancy of a locale and the continuity of localized social relations. Instead, we have seen the situatedness of the extended milieu shift towards 'significant places', 'decontextualized knowledge' and 'general-ized milieux'. These alternative forms of anchoring a milieu indicate what shall be the next topic of concern – aspects of the re-embedding processes. Giddens's notion of 'disembedding', as already mentioned, does not only imply the 'lifting out' of social relations, but, and just as important, their 're-embedding' (1994: 21, 87).

Leaving aside here the emphasis on 'abstract systems' and 'face work commit-ments' in Giddens's argument, I shall draw on the other, rather implicit aspect – namely the value-relatedness of these 're-embedded contexts of interaction'

81

(1994: 87). In continuation of the argument developed so far, it can be claimed that the extended milieu also finds situatedness in association with the like-minded. Moreover, the maintenance of the extended milieu's situatedness has so far been described as a fairly individualistic project. The following section serves to illustrate that the disruption of local milieux, and the subsequent individualization and extension of milieux, do not necessarily spell social individualism or subjectivism, but instead open up the possibility for new forms of social engagement.

The idea of an *association of the like-minded* can call upon Scheler's investigation into the different forms of social relatedness and their respective forms of communal experience. Important to note in Scheler's argument is the emphasis placed on shared values as the foundation of communal experience. This, according to Scheler, applies to a variety of social units, from the lasting 'life-communities' of family and tribe to the more sporadic communal experience of the 'mass'. What is specific about the third form of social unity, 'association', is the fact that it is a deliberate engagement between the individual 'self' and 'others'. Belonging to and participation in an 'association' in Scheler's sense is subsequently a question of personal commitment rather then birth or local attachment (Scheler 1973: 557; see also Dunlop 1991: 34). More recently, this idea of people's social engagement on the basis of individual choice and aesthetic reflexivity has surfaced in Lash and Urry's notion of 'invented' or 'new communities' (1994: 50, 316) and Maffesoli's concept of 'pseudo-tribes' seen as 'emotional communities' of taste and lifestyle (1996: X, 6ff.). Important for our argument is that these associations of like-mindedness provide a frame of action and meaning that is neither territorially bound, nor does it subscribe to the primacy of local social relationships over those maintained across distance. Instead, these forms of association are primarily mediated through overlapping and disjunctive 'fields of effectiveness' constituted by shared values (Scheler 1973: 557), shared symbols (Lash and Urry 1994: 50) or the 'aura' of certain forms of solidarity (Maffesoli 1996: 18).

Globalization processes have certainly further encouraged the generation and maintenance of social environments based on like-mindedness in the above sense. As mentioned earlier on, the 'time–space compression' (Harvey 1993) brought about by global means of transport and communication has radically changed the geography of 'near' and 'far'. Proximity becomes a category relational to the individual's milieu. The 'like-minded' and therefore 'dearest' may live elsewhere in the world, and yet are effectively 'nearer' than the person that happens to live 'next' to us in spatial proximity.

This point can be illustrated by recalling some of Barbara's narrative. She keeps in touch with friends and family in many different places. These contacts are more important to her than close contacts to neighbours in the road on which she lives in Streatham. While in two years time she will move on from this more or less accidental place of residence, her networks of friends and relatives, which stretch from Nashville and Mexico to Hong Kong and Manila, will remain, and perhaps be extended via lasting contacts in London. However, these London contacts are likely to be those made in Filipino London where most of her socializing takes place.

Similarly, Ulla's narrative revealed the importance of her links to friends in Newfoundland in order to cope better with crisis situations in her milieu. The fact that even at the age of eightysomething, Ulla still regularly goes through the considerable inconvenience of transatlantic travel, suggests that for her, these links across distance are at least as important as the neighbourly contacts in Streatham.

We have to note here that Hannerz is certainly right to argue that global mobility is not confined to transnational cultures and people with a cosmopolitan outlook (1992: 238). Yet, while global means of transport and communication become more and more accessible potentially for all of us, we would at the same time have to observe at least an uneven distribution of that access. Associating with the like-minded across distance requires not just shared values and individual commitment, but also access to shared means of transport and communication; we are, in fact, talking about the life-world of privileged milieux.

However, we do not necessarily have to be globally mobile to have access to the internet, or to associate with the like-minded. The reverse side of global mobility is recognizable in the globalized world city itself. In this environment of micro-globalization, with the presence of people from literally every corner of the world, from a variety of ethnic, religious and social backgrounds, and with so many different interests, hobbies, and life styles, it is unlikely for anybody not to find the company of a like-minded soul. Thus, the globalized world city sees the rather spontaneous generation of associations of the like-minded, based on shared values and overlapping practical interests.

In this context, then, we might see the development of a new type of *non-localized social milieux*. These, in distinction from communities, emerge wherever individual milieux overlap to such an extent in their value disposition and their practical interests that these materialize in shared social practices and routines. In this definition the individual's milieu can be part of many social milieux without being completely absorbed by them. To illustrate, let's look again at Barbara's story. Due to her strong bond to the Philippines, she attached herself to a Filipino family in London, and subsequently got to know other Filipinos through this family. Also, we may recall, whenever and wherever London's Filipino community has a festivity, Barbara makes sure she is part of it. Her own account clarifies that she values this attachment for the continuation of communal experience that stretches back into her past, and therefore gives stability to her actual milieu in London.

> With the Filipinos I must say, I also *feel at home with them* [my emphasis], because, after having lived in the Philippines for so long. So for instance, with the Ferreiras, I was with them the whole day yesterday. We were

looking for familiar places, and so we went to Chinatown yesterday. And
when I was going to the British Museum, the girls of the Ferreiras were
accompanying me there. I feel a little bit more familiar with them than with
other nationalities.

The 'Filipino experience', as an organizing frame of Barbara's London milieu, also
extends towards work. Amongst the many social contacts she makes as a member
of the organizing committee of a world congress on women and media, Barbara
associates mainly with four women, who also have some kind of link to the
Philippines. They visit each other for meals, go to exhibitions and organize weekend
trips to the environs of London. Barbara describes this group of women, of whom
only one is American, as 'a setting for feeling really quite well, since we all have for
one or the other reason a history with the Philippines'.

It has to be stressed at this point that this example of an association of the like-
minded does not evolve around a locale. Instead it is the Filipino experience in its
various manifestations that provides the anchoring of this overlapping network
of like-minded people. This is not to say that associations of like-mindedness
might not centre around a specific locale. What comes back into consideration here
is a point made earlier regarding the role of 'significant places' within extended
milieux. Similarly, we could argue at this point that a social milieu of the like-
minded can evolve around a significant local setting without this setting gaining
primacy and turning the symbolic, that is value-based, association of the like-minded
into a local community. What we encounter here can be regarded as yet another
aspect of what earlier was described as the 'delinking' of locale and milieu.

We can illustrate this point by looking at the Lutheran Church of St Anne and
St Agnes in the City of London as one such locale around which a social milieu of
like-mindedness evolves. The 'Guide to Worship in Central London' (Willows
1988: 177) aptly introduces this church as 'unique amongst City churches in that it
is used for Lutheran worship (in three different languages), and also remarkable for
the variety of music that can be heard here'. What is remarkable with regard to the
delinking of local and milieu is the fact that St Anne and St Agnes is co-inhabited by
two different social milieux. The church was initially used exclusively by Estonian
and Latvian Lutherans in London, until in 1966 they invited the English-speaking
Lutheran congregation of St John's to share the site with them. Before that the
congregation of St John's moved between temporary premises. Although sharing
the site of St Anne and St Agnes, the St John's congregation maintained its separate
designation until 1988, thus being named St John's Lutheran Church *at* St Anne and
St Agnes' [my emphasis] ('Lutherans in London' May 1991; Tull 1974: 13). The
careful naming indicates a conscious differentiation between congregation and church
building, that is, between milieu and locale. The arrangement between congregation
and host locale emphasizes the primacy of shared values, leaving the actual church

building in a position of secondary importance for the continuity of 'St John's Lutheran Church'.

The delinking of locale and social milieu comes further to the fore when looking at the actual arrangements of the two congregations that share the church. There is certainly enough overlap in values and practical relevances between the two Lutheran congregations to make worshipping in the same church both acceptable and practicable. And yet, despite sharing a basic structure of values, in their wider structure of relevances these two congregations are so different that they constitute two different social milieux that hardly intermingle at all in the locale they share. There is a clear time–space stratification between these milieux. While the English-speaking St John's congregation worships on Sunday mornings, the Estonian and Latvian congregation meets on Sunday afternoons. Due to the historic developments before 1989 the Estonian and Latvian congregation still carries the relevances of a rather closed milieu of 'exiles' (Hannerz 1992: 242), attempting to provide a symbolic home to those forced to live spatially disconnected from their Baltic home region. The inward-looking character of this milieu is symbolically emphasized by the use of Latvian and Estonian language for worshipping, thus excluding other Lutherans.

In contrast, the English-speaking congregation attracts long- and short-term 'expatriates' living in London from all over the world, making its services a meeting place for expatriates not just of Lutheran belief. Pastor Englund actively encourages the social dimension of this congregation and believes that it contributes to the attractiveness of St Anne and St Agnes.

> I'm thinking just recently of three young women who are au pairs, one from Madagascar, one from Hungary, and one from Slovakia. Now, they only met here, but they come to our church, and they've brought friends. Now, I didn't count their friends, but these are three people that come every week now . . . And I think it's important for them to come and to meet other friends. And what makes me very happy is when I see people that meet at the church going off together, you know, and they may be going out for a meal, or they may arrange to go to a concert or a play together, and I see them exchanging phone numbers, and that makes me very happy. Because London, for many people can be very lonely place, and bewildering . . . So in a small way I would like our church to be, and that's what I think we do as best as we can, to provide a sense of community, and welcome the stranger, and that's a biblical theme that I think is important in our setting, it wouldn't be true of every church.

The fact that English, which for many people across the world is their second language, is used for worshipping and socializing, adds to the world openness of this social milieu, and making it easier 'to welcome the stranger'. This can be further illustrated through the congregation's presentation to the public. In one of the monthly church letters ('Lutherans in London' October 1991), for example, it says:

> God has blessed us with many people from all over the world and, at any given time, our membership is represented by twenty two different countries.

The world-openness of the social milieu at St Anne and St Agnes can be further illustrated by looking at some figures[1] concerning its membership. The English-speaking congregation of the church integrates 507 people from 39 countries. The largest national groups are:

UK	106 people	or	21	per cent
Tanzania	83 people	or	16	per cent
USA	80 people	or	15	per cent
Hong Kong	60 people	or	12	per cent
Ethiopia	45 people	or	9	per cent
Germany	24 people	or	5	per cent

It is interesting to note that these figures do not just include baptized members of the Lutheran church, but 'friends' of St Anne and St Agnes, a term used for those who participate in a service at least once a month. This little side note becomes rather important when raising the question of what is it that actually attracts people to this particular church in the middle of the City of London? For many it is certainly the wish to worship in the Lutheran tradition. But for others it is the fact that in this setting they encounter like-minded people in the wider sense. Here, expatriates from all walks of life, from high-flying executives to au pairs, come together in a sociable atmosphere. From the perspective of a long-time participant observer, it seemed that the socializing around coffee and cookies afterwards is at least as important as the actual service. These people share a similar 'milieu structure', in other words, live a more or less internationalized and fragmented family life, share a more or less cosmopolitan life experience, share the experience of temporary employment and short-term postings, and are faced with the task of making London temporarily their home. It is these coffee sessions where the telephone number of a good dentist is passed on, where the au pair meets a family in need of an au pair, where a newly-arrived member becomes acquainted with someone who knows someone else who might rent out just the type of flat (s)he is looking for, and so forth. In other words, participation in the 'milieu of like-minded' at St Anne and St Agnes is not primarily or exclusively about shared religious values, but based on overlapping practical relevances of living the life of an expatriate in London.

We should not be surprised to meet Barbara again in this setting. Barbara occasionally goes to the Sunday service at St Anne and St Agnes not so much for reasons of worshipping, for she is a Methodist, but in order to meet other Americans and socialize with other expatriates. She likes the relaxed atmosphere during the coffee sessions after service. What attracts her to this setting is both the variety of people she meets, and the possibility to associate with the like-minded. Next to the Filipino connection she considers the St Anne and St Agnes setting as her home in London.

> I found people in that setting that I like to associate with. Going to St Anne and St Agnes you meet people from all over, and you get acquainted with them quite easily, it's a very friendly congregation, because they're all trying to find their way in a big city.

But it is more than spiritual support that Barbara gets from this setting of like-minded people. Especially in her settling-in period in London, it was here that she would find the people with the relevant resources of information and practical advice.

> So I think the 'resource people' you find, I mean going to St Anne and St Agnes, you meet people there who know how to take care of things, or how to get round to places you want to see in London, and that I found quite helpful.

Barbara's account clearly illustrates that the attractiveness of this social setting goes beyond like-mindedness in the purely religious sense, and instead extends into the practicalities of expatriate everyday life in London. This observation is also stressed by Pastor Englund. According to him, for most members of the congregation, it isn't so much the intention to meet exactly the same people week in week out, that brings them from places all over London. Instead, it is the chance to meet people who each have a unique biography, and yet share a similar enough background with regard to religious beliefs, social and intellectual background, and problems they have to deal with in London. He believes it is this formula of very different and yet similar people that makes St Anne and St Agnes such a pleasant social setting.

> What they do meet is, they meet people from thirtysomething different countries, a number of whom are from their particular [religious] tradition, and that's interesting, because there are *similarities and yet differences* [my emphasis].

In conceptual terms we could say that St Anne and St Agnes provides a meeting place for people with unique 'biographical situations' (Schütz 1970), and yet a sufficiently similar or overlapping 'structure of relevances', which makes socializing relatively easy. People whose milieu more or less carries the index of an 'expatriate', find

87

relevant social contact here without having to engage in close-knit social relationships with particular others. From this point of view the social setting evolving around St Anne and St Agnes could then also be described as an *alternative support network* for those who are separated from the traditional support networks of local community and extended family (Beck 1992: 93), a scenario certainly familiar to many expatriates.

It is exactly for the reason of emotional and practical support that Harold, the 'global business man' values the St Anne and St Agnes setting. He participates in it mainly through his wife, who organizes the London milieu of the family while Harold is somewhere between Europe and New York. While other expatriates might be integrated in the social life evolving around the corporate structure, Harold's 'pigeonhole office' does not provide such a social setting.

> Because, you see, in my corporation I'm the only expatriate American in the headquarters. In other corporations, like the big oil companies or the big banks, there is generally twenty or thirty families, you know, American families. So certainly, if you are new and you come here, you have an immediate liaison to somebody. So then they are in, I mean their wives are in where the wives are, the husbands are where the men are, you know, they immediately know the golf courses, the shooting courses and everything else.

As Harold and his family also do not live in one of London's residential enclaves dominated by expatriates, as for instance Holland Park or St John's Wood, their social contacts mainly developed through alternative settings of like-minded people with some sort of shared interest.

> The development of social contacts was mainly through the children, naturally, because they both go to the American School. Children are important. It's an easy thing to get on to, you know. You can go to the school, you have your children there, there are a lot of other people who have their children there, and so you immediately get together . . . But also the church, you tend to choose a church where there are Americans, or similar expatriates like you, OK. I shouldn't even say Americans, I say Americans because I'm American, but you tend to go to churches where there are expatriates, I mean socially you tend to do that, because *you are kind of looking for someone like you* [my emphasis], somebody who has the same problems that need to be discussed.

Both Barbara's and Harold's accounts of St Anne and St Agnes indicate that it provides the ideal setting for some sort of re-embedding of their rather mobile and extended milieux. The social milieu that evolves around this church is neither based on the constancy of the 'locale', as the expatriate congregation is only a 'guest' at this particular church building. But even its host, the Estonian and Latvian congregation does not own the place.[2] Neither is this milieu based on the continuity of social ties between particular individuals. There is no 'in group', in other words, as there is generally a considerable turnover in this congregation, and also a big fluctuation of attendants from week to week.

While normally transience would be regarded as counterproductive to the generation of a community spirit, this seems not to be the case in this particular setting. It turns out to be this rather transient character of the congregation and the ephemerality of social contacts that assists this particular association of like-minded people. Pastor Englund believes that this turnover actually provides the internal stability of the social milieu evolving around his church.

> I think the turnover helps. Many other churches . . . have a majority of people who have been there a long time. And, one of the most difficult experiences is a group, a congregation that looks on itself as a kind of family, an in group, and then the visitors come, and it takes them a long time to break in, or they're kind of rebuffed, they're looked upon as threat, whereas everybody is a visitor at St Annes . . . I think it is much easier with this turnover and with a congregation where everybody is different.

This seemingly contradictory statement – turnover generating the situatedness of a social group – reveals yet another implication of the delinking of locale and milieu. The transitory character of the expatriate life style, plus the genuine search for the like-minded in the light of lacking the traditional support networks, make it, in this case, impossible to generate 'established/outsider' constellations. In conceptual terms, the transitory social milieu of the expatriate congregation at St Anne and St Agnes is emancipated from a form of social situatedness that is defined entirely through the long-term settledness of a community in a particular place (Elias and Scotson 1965: 2f.).[3] The mere length of residence or presence in a particular place as a pure value for the cohesion of social relations is in this case replaced by situatedness based on like-mindedness in action.

Similar to the function of 'significant places' in the context of the individual's extended milieu, it could be argued that social milieux of like-mindedness find attachment to particular locales, but are not defined by them. It is people who generate the 'aura' and relevance of significant places. Pastor Englund, for example, is quite aware of the fact that such a social setting as has developed at St Anne and St Agnes is kept together by an atmosphere that is generated from inside the place by the participants themselves. Moreover, in this particular case, social situatedness is further encouraged by the charisma of significant figures who set standards through their own attitudes, their own 'world of values' in action.

> Then again, it's partly my personality and Ruth's [his wife] personality, I think. We really love people, and love to meet new people and find it exciting, and I mean, what I really believe, that it's a privilege for us to welcome new people. We don't look at them as strangers, we are honoured to have them come, that's very exciting to me, and we learned so much from one another interculturally, I mean that's very important.

What comes also back into consideration here, is the interplay between social environment and individual milieu. The welcoming and open atmosphere at St Anne and St Agnes can only provide a potential setting for the re-embedding of the milieux of people such as Barbara and Harold. How much people feel at ease in this social setting characterized by similarity and yet difference, depends on the degree to which the individual's milieu shares the value structure manifested in this association of the like-minded, such as general open-mindedness, an outgoing attitude towards the newcomer, and tolerance towards people's ethnic and political background. This openness towards experiencing ethnic and cultural variety in the context of shared religious belief does not appeal to everybody just because (s)he is an expatriate, as Pastor Englund has had to acknowledge over the years.

> We have some American families, for example, they come, and they feel threatened by this, and they would subsequently rather go to the American church, which is largely just like being at home, you know, back in the States. Whereas other find it just an eye opening experience.

The association with the like-minded, however temporary and transient, is a voluntary engagement and can be withdrawn. So whoever engages in the group does so rather wholeheartedly, beyond the normative pressure of an established in group. The potential re-embedding of individual milieux in social milieux of like-mindedness thus brings to full consequence the process discussed at the beginning of this chapter: the 'uprooting' of milieux and the 'delinking' of locale and milieu. The association with the like-minded in its logical consequence means the realization of the milieu as a value-based configuration of action and meaning, set free from the social and normative primacy of localized frames of action and meaning.

6

STREATHAM – THE RELUCTANT
SUBURB: THE METROPOLIS
EXTENDS

The so called romantic character of any particular spot arises from
a calm feeling of the sublime in form of the past, or of solitude,
absence, or seclusion, which comes to the same thing.

J.W. Goethe (n.d.: 170)

Sitting on a Sunday evening on one of the benches overlooking a little pond in
the Tooting Bec Common, listening to the bells of Streatham's St Leonard's Parish
Church calling for the 6.30 evening service, and watching the elderly people on the
other benches reading their books and newspapers, while all of us enjoy the first mild
summer's evening, the participant observer could be forgiven for assuming he was
in the middle of the tranquil idyll of an English village. It is only when this tranquillity
is disrupted by the siren of a speeding police car, a scenario resembling a badly
produced television crime serial located in New York or Los Angeles, that the
realization dawns that one has been caught up in an illusionary idyll. With the police
car coming in sight and having 'Metropolitan Police' written all over it, one recalls,
we are right in the middle of London here, the latent threats and the hectic rhythms
of the world city did not stop at Streatham's borders. It is this interplay between
parochial tranquillity and metropolitan rush that actually generates the unique
character of many of the villages and small towns that have been swallowed by the
extending metropolis, and which try to hang on to some kind of local identity.

It is not just the scenery of some of the locality that radiates an idyllic aura. What
is intriguing for the newcomer, revelling, perhaps somewhat naively, in the sheer
speed and transitoriness of London life, is the nostalgic romanticism about the place
that is readily expressed by a substantial number of local residents and rather
forcefully pursued by the two resident associations. But what looks at first sight
like a contradiction, residual parochialism in the middle of the global city, becomes
less of a paradox when looked at against the background of Harvey's argument
concerning 'time–space compression'(1993: 240ff., 284ff.).

In general terms, time–space compression opens up a paradox of place and space. Historically this process means the collapse of spatial barriers and with it the incorporation of localities into 'homogenized space', generated by a universal and comparatively standardized system of transport and communication and the capitalist commodification of space. This, according to Harvey, implies the end of the relative autonomy of place within space, and the disruption of the relatively isolated social worlds evolving around place. As places become increasingly open and vulnerable to outside influences, this can mean an undermining of local identity and a severing of the link between place and personal, as well as collective identities. But, and herein lies the paradox, this incorporation of place into global space increases at the same time the significance of places in that relative local advantages come to the fore more clearly and can be actively promoted on a global scale (ibid.: 1993: 293ff., 303ff.).

Harvey points out that, while place keeps it significance within the new geography of time–space compression, the meaning of place within these circumstances has changed. Rather than being defined through its geographical and local authenticity, place becomes the 'intense focus of discursive activity' (Harvey 1996: 316f.). In turn, this 'fashioning of some localized aesthetic image' produces personal and collective place identities that are best described as 'affective loyalties' (Harvey 1993: 303f.; 1996: 323). In this sense, then, place is not a given, but something that has to be actively produced and maintained in relation to other localities. Yet by no means is this construction of place an arbitrary enterprise. While on the one hand the imaginary identity of a place will be charged with 'all manner of personal and collective hopes and fears', it is the 'genius loci' – the memories and meanings gathered by a place throughout history, for example in its buildings – which provide the playing field for the invention of local identity (Harvey 1996: 321, 308). From this perspective it makes sense to argue that any place-bound identity and any search for local roots 'has to rest at some point on the motivational power of tradition' (Harvey 1993: 303).

The argument developed by Harvey helps to shed light onto the role of tradition and local identity in the specific case of time–space compression taking place within the metropolis turning globalized world city. During the nineteenth century the metropolis expanded steadily towards its hinterland. In its most intense period, in the second half of that century, places like Streatham or Peckham, that were once towns and villages in their own right, were swallowed by the metropolis. The incorporation into London's built-up environment and the metropolitan system of public transport, ended the relative autonomy of these places and their local communities. From then onwards we see the discursive attempt to maintain the permanence of a place called Streatham within the rather homogenous and yet ephemeral urban landscape of the world city. The turn towards images of Streatham's past provides the quasi-stability for those searching for roots in a shifting metropolitan landscape, however aesthetisized and mediated they might be. We should not immediately equate this aesthetization of locality with a revival of local community. It is often those whose milieux are mobile and extend beyond a place that are most actively engaged in the production of place in the above sense. While the local discourse might serve to achieve some level of temporary local solidarity, it is not equipped to integrate milieux coexisting within a place into the intimacy of a community of face-to-face interaction.

This chapter will illustrate further some of the issues concerning metropolitan time–space compression by engaging in more detail with the past and present of the Inner London suburb of Streatham. The case study of the historic links between London and Streatham will also serve to grasp the historic dimension of the delinking of local and milieu, which has been missing from the argument so far.

Streatham – the would-be West End of South London

Located in the South West of London, Streatham is the most south western part of the borough of Lambeth; the North of the borough directly borders the River Thames. The area contains two substantial parks, Tooting Bec and Streatham Commons, both remnants of the Surrey countryside that existed before the sub-urbanization of South London. Today Streatham is by definition part of Inner London. It is just inside the six-mile radius from Charing Cross, the supposed geographical centre of London. Yet the newcomer might find that London's famous black cabs refuse to transfer him or her to Streatham, as they are reluctant to go 'South of the River' in general, and particularly reluctant to go through Brixton, London's version of the Bronx.

Streatham has a population of 50,000 people and an unemployment rate of approximately 20 per cent, which is in keeping with the figures of other Inner London boroughs (Lambeth Unitary Development Plan 1992: 225). In its ethnic mix Streatham reflects the general trend of 'all of Greater London's boroughs [becoming] more cosmopolitan between 1971 and 1981' (King 1991: 141). Streatham differs slightly, though, from its neighbours Tooting and Brixton, where the Asian and Caribbean culture respectively give a certain atmosphere to those neighbourhoods, in that no ethnic community visibly dominates its busy street life. Streatham really is a mix of nationalities from across the world. This is displayed for instance in the truly international cuisine assembled along its High Street, including Italian, Spanish, Malaysian, Chinese, Greek, Turkish, and Indian restaurants, but also if one ventures into any of the more quiet side streets, where one can find, for example, 'Mario's', the Greek-Cypriot barber, next to an Italian restaurant and a Polish-run secondhand book shop. And if the evidence of 'naturally occurring settings' (Silverman 1989: 156) is anything to go by, then Preston Road,[1] a compara-tively small residential road off Streatham High Road, gives further illustration of this truly cosmopolitan mix. Here, we find the English family living next to an American couple, an American-German couple, a Sri Lankan GP, an old Polish man, an English retired elderly lady of Swedish background, a Cypriot-English couple, an Italian-English couple, a young Frenchman, and last but not least, the Asian newsagent from Tanzania.

Streatham High Road contains most of the retailers normally present in any Inner London High Street, such as W. H. Smith (stationery and office supplies), Dixons (computer and Hi-Fi), Our Price (CDs and records) and Woolworths (general domestic consumption). There is also the normal selection of super-markets such as Safeway, Sainsbury's and Iceland. It has got its fair share of pubs and restaurants, as well as takeaways, competing for customers with a rather big branch of McDonalds, as well as KFC and Pizza Hut outlets. Finally there are branches of almost all national banks, from NatWest to Barclays, complemented by the usual set of building societies and travel agents. Some special character and style is added to Streatham High Road by the old St Leonard's Parish Church, the church tower of which is some landmark and gives the flair of an old market town. This impression is also further given by the building of the Tate Library, though behind its nice facade metropolitan reality has moved in, with the unemployed and homeless making the best of the reading room shelter that combines the pleasure of a warm place with the convenience of free access to newspapers and books.

Streatham's built-up environment largely consists of venerable Victorian and early Edwardian Houses, occasionally interspersed with modern estates, partly due to the damage done to Streatham during the Second World War. In particular, the neatly lined up Victorian semi-detached properties of the residential side streets off the High Road, in which the milkman still delivers to people's doorsteps, still indicate that today's Streatham is largely a product of the Victorian suburbanization of South London, facilitated by a hugely extended railway network to which Streatham was linked up between 1856 and 1868 (Gower 1990: 22). Yet the fact that many of these elegant Victorian houses are no longer single properties but divided into flats, again, might serve as an indication of Streatham's ambivalent identity. While the architectural setting still indicates the rather elegant character of pre-war Streatham, the social problems that come to the surface, also through the beggars and homeless people sheltering in doorways and selling the *Big Issue* outside supermarkets, are a reminder that Streatham is among the poorer Inner London boroughs.

There is another fine distinction concerning Streatham's identity in relation to its neighbours Balham, Tooting and Brixton. Streatham, as opposed to the adjoin-ing areas, is not served by the ultimate means of metropolitan transport, the 'tube' (London Underground). Instead it is served by surface rail operators. From Streatham Hill Station and Streatham Common Station one can catch a train to Victoria every 15 minutes, with the journey taking about 20 minutes. From

Streatham Station trains to London Bridge depart every 30 minutes, and the journey, again, will take about 20 minutes. Streatham is also quite well served by London buses. Several routes link Streatham directly with Central London, the 159 bus, for example, goes directly to Piccadilly Circus, Oxford Circus and Baker Street. For the uninitiated newcomer this density of metropolitan transport seems quite impressive. Yet, as people in Streatham are quick to point out, the issue here is not so much a technical one, though frequency of trains is a problem. What is at stake here is more the symbolic image of the place; Streatham is crucially missing from any tube map – the essential frame of orientation for Londoners and tourists alike. Streatham in this sense is 'off the map' as concerned locals will argue. This exclusion from the network of London Underground *de facto* makes Streatham an Inner London suburb, in other words turning it into 'just another place name, part and parcel of that somewhat anonymous area known as South London' (Gower 1990: Introduction).

It is this image of Streatham as just another part of London's suburbia that does not sit too well with its residents, in particular those rallying around the two resident associations, the 'Streatham Society' and the 'Streatham Association'. They fancy Streatham as a place of fashionable reputation. This reputation is fed by a rather glamorous image of the Streatham of 'those days', reaching from the immediate past back to the medieval history of the locality. In this regard the division of labour between the two associations makes very much sense. While the 'Streatham Association' is more concerned with getting Streatham 'back on the map' of London, literally and metaphorically, the task of the 'Streatham Society' is to provide the historic backing for this campaign.

Leaving aside the historic dimension for the moment, even today's Streatham seems to lay justifiable claims to be the entertainment centre of South London. In the development plan of the borough council (Lambeth University Development Plan 1992: 229) it says in this regard:

> Streatham is an important South London entertainment area. It is unusual in that it contains a wide range of leisure facilities in a relatively small area. Streatham Ice Rink is one of only three such facilities in London and draws people from all over South London. 'The Ritzy' dance venue is a similar South London attraction. The Streatham Mega Bowl is one of South London's few bowling alleys. In addition, there are two multi-screen cinemas, a swimming pool, a bingo hall (formerly the Streatham Hill Theatre, able to seat over 2,500 patrons), numerous restaurants and wine bars and clubs for tennis, bowls and squash.

Especially on a Friday or Saturday night, when people flow in and out of the numerous pubs, queue in front of the Mega Bowl complex, and crowds gather in front of the

'Odeon' and 'Canon' cinemas, it feels like one is in an outlet of London's West End. Judging from these impressions the newcomer is ready to accept the locals' vision of Streatham being the 'West End of South London', with Streatham High Road figuring as the 'Bond Street of the South'.

A rather ambivalent note to Streatham's reputation as an entertainment centre is finally added by the prostitutes and the related kerb-crawling in the side streets off Streatham High Road. The prostitutes clearly see the relative advantage of place, with Streatham being within easy access from King's Cross, London's alleged centre of the sex trade, and attracting the 'leisure seeking' from across South London, while at the same time it provides the comparatively safe environment of a respectable neighbourhood. Local residents are obviously more ambivalent about this aspect of Streatham's night life.

So overall, despite a few boarded up shop windows, traffic congestion and pollution, and the lack of direct access to London Underground, the newcomer cannot help but see Streatham as a lively and entertaining place, a sought after place to live in. It is therefore not immediately accessible why local residents should be concerned about Streatham's future, and should complain about the decline of the neighbourhood; 'Streatham is going downhill' is the most common expression used by young and old alike. Asked what precisely they mean by 'the area's decline' (Streatham Association 1993), people mention the overall impression of the place in comparison to their neighbours Balham and Tooting, to which places Streatham has supposedly lost out. The curious newcomer will then learn that behind this comparison is another sore point that indicates the complex problem of place within a shifting metropolitan landscape, with history playing a big part. Streatham became part of the borough of Lambeth only in 1965 as a result of an adminis-trative shake up of Greater London. Before that it was part of the rather affluent borough of Wandsworth, like Balham and Tooting still are. Not surprisingly, then, it is one of the campaigning issues of the resident associations to reverse this disruption of a historically-developed pattern of belonging by a bureaucratic decision.

There are then further issues that are brought forward as indicators of Streatham's decline. To begin with, there has been a subtle change of character in the High Street following the 1992–93 recession. Fashionable retailers have closed, partly being replaced by cut price retailers, charity shops and building societies, giving the High Street a rather ordinary character, apart from its entertainment facilities. Not to mention the occasional boarded up shop window, allegedly something unknown some 10 years ago. As were people rummaging through litter bins, now a common picture, just as in other parts of London. Residents are also concerned about the loss of Streatham's reputation as a safe area, underlined, for example, by the following incident described in a police notice:

Police Appeal for Assistance
£5000 Reward

Police are treating as murder the death of
Mrs Constance Brown, 72, of Streatham,
after she was robbed and seriously assaulted
in the forecourt of Streatham Bus Garage,
Streatham High Road, at 3.40pm on Friday,
2 April 1993.

Then there is also the aforementioned problem of Streatham not being connected to the network of London Underground – a long-standing campaign issue in local politics. However, the most obvious loss to Streatham's High Road is the closure of the Pratt's department store in 1990. The partly boarded up, partly visible ruin of the department store has developed into the symbol around which the local political discourse evolves and local identity is amalgamated. What is at stake here is not just the loss of a convenient shopping facility for local residents, however important this might be for people like the elderly lady who said in one of the meetings of the Streatham Association, when potential redevelopment schemes were on the agenda, 'We don't want little shops, we want a department store back. We don't want to go up to Oxford Street for a pair of knickers'. But apart from being part of the rather prestigious John Lewis chain, Pratt's used to be one of the landmarks of Streatham. For people coming along the A23 approaching Streatham from the South, the rather impressive building they faced when turning into Streatham High Road would make it distinguishable from other High Streets along the way towards Central London. More importantly in the context of space and place within the metropolitan urban landscape, the department store also gained Streatham a name beyond the immediate locality, attracting shoppers from the adjoining boroughs (Lambeth Unitary Development Plan 1992: 225).

Given this real and symbolic significance of that department store, Lambeth Borough Council duly regards its closure to be a 'major loss to the area' and admits that there is 'the need to find a replacement quality retailer for the site as soon as possible' since 'there is the demand for better quality shopping in Streatham' which makes 'the redevelopment of the Pratt's store so vital to Streatham' (ibid.: 226). Despite this acknowledgement, throughout the 1990s Pratt's remained a ruin, symbolizing the 'decline of the area', and accordingly has become the most symbolically charged site on which the struggle for local identity focuses.

There is a further historic significance to this closure, especially for the older generations of residents, as it is the last of several more glamorous sites that were closed or replaced by low key entertainment, steadily eroding Streatham's name as the 'West End of South London': the Streatham Hill Theatre, closed in 1962, was redeveloped as a Bingo Hall; the Locarno Dance Hall, also closed in 1962, was redeveloped as the Ritzy Disco, later to be replaced by Cesar's Palace; the Gaumont Palace Cinema was redeveloped as a bowling alley. It is the glamorous reputation of the Streatham of 'those days', illustrated and backed up by a considerable amount of local history studies, which is used by the Streatham Association to bargain for a better future of the place based on the standards and merits of a bygone past. And this bygone past stretches much further back than the period before and after the two world wars. Much of the local identity discourse focuses on the time when Streatham was a town in its own right. We therefore ought to have a closer look at Streatham's history in order to discover the changes that came about for the place with its incorporation into the expanding metropolis.

Streatham High Road – early links between Streatham and London

First documentary evidence of Streatham goes back to the Chertsey Register of 727, produced by the Benedictine monks of Chertsey, which supposedly mentions an estate at Totinge cum Stretham. The second documented record can be found in the Domesday Book of 1086, compiled by Norman clerks after the Norman conquest in 1066 (Bromhead 1936: 7f.).

Streatham's early history is above all the history of what is today its High Road. Its origins allegedly go back to a small settlement built during the construction of a Roman road that linked London with the Sussex coast near Brighton. It was then the Saxons succeeding the Romans who gave the settlement its name, in its original meaning linking Streatham with the nearby road: 'the dwelling by the street' (Weinreb and Hibbert 1983: 833; Gower 1990: 1ff.).

So if anything, Streatham's early history is an example of the relational character of place, as from the outset it was influenced by being in the wider catchment area of London, conveniently placed on a major thoroughfare to and from the metropolis. Developing in one of the resting places for travellers, Streatham's development in the Middle Ages culminates in it being granted 'the right to hold a fair and a market' (Weinreb and Hibbert 1983: 834). The improvement in communications in eighteenth-century England (see Trevelyan 1946: 380ff.) is reflected in the fact that the road connecting London and Brighton becomes an authorized turnpike road in 1717. Streatham again benefits from its location when the Horse and Groom Inn, still in use as a fashionable pub near Streatham Hill Station, becomes 'one of the official stops for the stage coaches'. The Prince of Wales, later to be George IV, often used the place as a stopover on his way to Brighton, and the Horse and Groom Inn 'gained a reputation for gambling and cock-fighting' (Weinreb and Hibbert 1983: 834; Gower 1990: 18).

The above features of Streatham's history can be seen as what, following Harvey (1993: 294), could be described as deriving from its 'relative locational advantage'. But there were other circumstances, partly but not wholly accidental, that linked London and Streatham. In 1659 a discovery was made 'that was to place Streatham on the map' of the London society: the discovery of the Streatham Wells. Soon a spa was developed around the wells, the waters of which 'were found to have medical properties with a "mawkish taste" and were declared good for worms and for the eyes' (Weinreb and Hibbert 1983: 834f.; Gower 1990: 14). Streatham became a major attraction for Londoners. When the spa was at the peak of its reputation at the beginning of the eighteenth century, concerts were held twice a week. It is said that 'Streatham High Road and Common were then fashionable promenades, where you might meet all the well known leaders of fashion and society' (Arnold 1886: 99).

But it wasn't just London society who flocked to Streatham; its name was actually established within London's emerging Coffee House Culture via the fashionable 'Streatham Waters'. In a newspaper advert from 1717 it says:

> The true Streatham Waters, fresh every morning, only at Child's Coffee-House, in St Paul's Churchyard; Nando's Coffee-House, near Temple Bar; The Garter Coffee-House, behind the Royal Exchange; The Salmon; and at the Two Black Boys in Stock's Market. Whoever buys it at any other place will be imposed upon.
>
> (cit. in Arnold 1886: 99)

It is against this background of its privileged position in relation to London-bound communications, and its reputation amongst well-off Londoners as a fashionable place to be seen, that Streatham began to attract the influential and wealthy, who found Streatham a convenient enough place of residence or attractive enough to have a 'country seat' in this little neighbour of London's. Amongst those who decided to take a residence in Streatham were people whose careers would undoubtedly be linked to the emerging world city, but who at the same time could afford to 'escape the hustle and bustle of London life' (Gower 1990: 12), including Richard DuCane, Director of the Bank of England; his son, Peter DuCane, a director of the East India Company; and John Hankey, merchant and trader with the West Indies. All three resided around Streatham Common. Memorials in Streatham's St Leonard's Parish Church give evidence of the membership of John Massingbird, the seventeenth-century Treasurer of the East India Company; and Edmund Tilney, Master of the Revels to Queen Elizabeth and King James I (Gower 1990: 12; Gower n.d.: 6, 11).

While this collection of names and professions already indicates a link between Streatham and the City of London, which was on its way to becoming a centre of the emerging world economy, it was the Howland family who actually made the name Streatham known beyond London. The family took residence in Streatham around 1600. As textile merchants, the Howlands obtained interests in the East India Company via martial ties with one of the company's key families. Later becoming important share holders and investors, the Howlands 'established Streatham's long association with the Company'. Important links with the locality were further underpinned with marriages into the local Duke of Bedford's dynasty by the Tavistock family (Gower 1990: 12; see also Borer 1977: 228). Streatham here reflects a social pattern that, according to Pahl (1973: 15), is significant for the beginning of the Industrial Revolution in England: the newly rich industrial merchant classes married into the local gentry in order to gain higher social status, and to combine success in the City with the pleasures of country life. Still today this constellation is reflected in Central London's street names, where we find the modest 'Streatham Street' in the neighbourhood of 'Bedford Square' and 'Tavistock Square'.

The link between Streatham and London, between local gentry and East India Company merchants, manifested itself in various ways. Borer (1977: 228) reports, for example, that at least one of the dry docks of the East India Company was named after the Howland family. Amongst the company vessels were the 'Howland', the 'Bedford', and the 'Tavistock'. But most significant is possibly the vessel 'Streatham', symbolizing the locality's relations with world trade via the Howland family's involvement in the East India Company. Between January 1747 and September 1750 the 'Streatham' carried the locality's name around the world, from London to Bombay and back, before the representative of the Duke of Bedford received the following letter:

> Sir,
> You are desired to meet the owners of the 'Streatham' on Friday, the 28th instant, at the Jerusalem Coffee House at one o'clock in the afternoon, to receive a dividend for His Grace the Duke of Bedford.
> I am Your most humble servant,
> [John Hallett.]
>
> (cit. in Borer 1977: 228)

What this little excursion into Streatham's early history certainly indicates is that the autonomy of place is always a relational concept, to be seen against the transitoriness of the wider socio-spatial environment. The Streatham of the Howlands was a comparatively self-sustained semi-rural country town, in contrast to the bustling Inner London suburb it is now. Yet at the same time we have early indications that the history of the place is not as autonomous and localized as the image of the Streatham of the past (cf. Loobey and Brown 1993) seems to suggest, around which the current discourse of its local identity is centred. Streatham's history, as looked at so far, rather provides illustration for Harvey's claim that 'all places are to some degree open' (1996: 310). Relatedly, 'Streatham Village' shows features of what Giddens calls a '*phantasmagoric place*' (1994: 108f.), or places that 'do not just express locally based practices and involvements but are shot through with much more distant influences'. This can be further illustrated by the fact that Streatham reflected on a small scale some of the increasingly international patterns of the 'Peopling of London', in particular reflecting its status as the 'world's asylum' (Merriman 1993: 20, 43). In the 1680s, after Charles II offered them asylum in England, a small colony of Huguenot silk weavers settled near Streatham Common. Similarly, after William III's access to the throne, Dutch families allegedly settled in Streatham (Weinreb and Hibbert 1983: 834). So while London increasingly influenced, and at the same time was influenced by worldwide links, Streatham is indirectly affected by these developments via its links with London society.

However, while Streatham's history shows it to have been a 'phantasmagoric place' from the beginning, this is not to say that 'Streatham Village' as a relatively autonomous and intimate local community has not existed. What we encounter here is the problem of social perspectivism, or milieu, to stick with our conceptual tool. The Streatham of the Howlands and Tavistocks was different from the Streatham of, say, the local barber and teacher. Different social strata generate different worlds of experience. The same locality can be a closed world for some, while for others it is a place from which to rule (at least parts of) the world. The following section will pursue further the implications of this problem with specific regard to early indications of a delinking of locale and milieu.

Dr Johnson's Streatham – early stratification of local and extended milieux

We have to recall here that time–space compression, in Harvey's notion of the term, is more than just a spatial phenomenon. The collapse of spatial barriers goes with the encounter of formerly separated social and cultural worlds, and the restructuring of the sociocultural landscape of society in the context of developments initiated with the Industrial Revolution (Harvey 1993: 238ff.). And this encounter of different social worlds is not confined to the distant and 'other' in geographic terms.

In the second half of the eighteenth century a unique constellation of social worlds finds its expression in Streatham's social microcosm, which reflects the changing sociocultural landscape of English society at the time of the accelerating Industrial Revolution. Trevelyan describes this society at the beginning of the Industrial Revolution as a stable and yet permeable society, a society of a new 'humanitarian spirit' in which economic success was complemented by a desire for art and learning

(1946: 339ff., 396ff.). While the aristocracy still functioned as patrons of art and education, it was joined by the new middle classes. Trevelyan (ibid.: 398) notes:

> The social aristocracy of that day included not only the great nobles but the squires, the wealthier clergy, and the cultivated middle class who consorted with them on familiar terms.

Vulliamy (1936: 20), in a study concerned with the actual constellation in Streatham, argues in a similar vein when stating that there was

> a certain confusion between the middle and the upper layer. The new grouping of capital under the industrial system, from 1760 onwards, naturally increased the permeability of the middle class, which was already preparing to join the elegant masquerade.

Apart from the upwardly mobile industrial middle class being prepared to finance a social life centred around art and learning in order to gain social recognition, there were two or three more elements that favoured this encounter of social layers.

First, the Act of 1751, which taxed spirits highly, ended the years of excessive gin-drinking in London between 1730 and 1750. Trevelyan (1946: 341ff.) considers this to be of such importance that he describes 1751 as 'a turning point in the social history of London', insofar as it encouraged a Tea and Coffee House culture, and with it the aforementioned 'new humanitarian spirit'. Vulliamy (1936: 25), in turn, regards 'the transition from the Age of Gin to the Age of Beer' important because it supported the permeability between different social worlds. For beer was now being drunk 'at the tables of well-to-do families, as well as in the servants' hall'.

Second, there was the already mentioned general improvement of the road network, leading to an increased mobility in English society, which again resulted in an 'high degree of social, commercial and intellectual intercourse' (Trevelyan 1946: 381ff.). Although this mobility focused a lot of social activities on London, which by then was the place to be for the nobility and the newly rich alike, two counter factors simultaneously improved the importance of smaller provincial towns around London. On the one hand not all wished to, or could afford a complete 'London Season' (Trevelyan 1946: 373). Consequently, taking residence in one of the provincial towns around London enabled them to still take part in London's social life, due to the improved means of transport. On the other hand, those wanting to take part in London life, but at the same time wishing to escape the 'hustle and bustle' of the metropolis, could opt for the same arrangement. Finally, while London became the focus of cultural life, this did not apply to the same extent to the Hanoverian Court, as allegedly 'George II patronized Händel's music but nothing else'. This meant that the patronage of art and learning had transferred to 'gentlemen's seats and provincial towns, each of them a focus of learning and taste' (Trevelyan 1946: 399).

Streatham now, as one of these semi-rural provincial towns, reflected this overall constellation, aptly named by Trevelyan as 'Dr Johnson's England' (ibid.: 371), in more than just a metaphorical sense, as the following example will illustrate. In 1740 the Thrales, a brewer family from Southwark, bought the Morfield estate from the Duke of Bedford, and built a Georgian mansion on it called 'Streatham Place' (Weinreb and Hibbert 1983: 834; Gower 1990: 15). As beneficiaries of the 'Age of Beer', with the turnover of their brewery multiplying, Hester and Henry Thrale were able to entertain a rich social and intellectual life at Streatham Place. Representing the new industrial middle class, the Thrales were 'anxious to obtain some of the material advantages, some of the purchasable advantages, of aristocracy; and they were particularly anxious to improve themselves by elegant reading' (Vulliamy 1936: 20f., 24f.). Among the celebrities entertained at Streatham Place were the philosopher Edmund Burke, the painter Joshua Reynolds, the biographer James Boswell, and last but not least, the poet and lexicographer Dr Samuel Johnson. All of them were regulars of the London 'Literary Club', and Johnson almost certainly something of a London character (see Weinreb and Hibbert 1983: 334; Gower 1990: 15). However, it was the friendship between Dr Johnson (1709–84) and the Thrales that allegedly 'has put Streatham and St Leonhard's Church for ever on the map of literature' (Bromhead 1936: 12). While the others were frequent visitors, Johnson worked and lived in Streatham on and off during the period between 1766 and 1781. He had his own room at Streatham Place, enjoyed the perfect hospitality of the Thrales, and gained inspiration during lengthy walks with his intellectual friends on Tooting Bec Common and Streatham Common (Bromhead 1936: 12; Gower 1990: 15; Weinreb and Hibbert 1983: 834). It is said that he wrote large parts of his main work, *Life of the Poets* at Streatham Place (Myers 1949: 131). Even today Dr Johnson Avenue, running through the Tooting Bec Common, is a reminder of that link between the London poet and the locality.

As stated before, in Streatham between 1766 and 1781 we find developing a constellation typical of English society at the beginning of the Industrial Revolution. Vulliamy (1936: 25f.) describes the setting a an 'obliteration of strata' in which there was 'nothing absurd in the idea of an alliance between brewing and elegance, or between literature and industry'. He points out:

> In the Streatham coterie . . . the different elements of the social pattern were clearly visible. Literature was grandly represented by Dr Johnson, who was also the massive embodiment of orthodox morals and of insular perversity. Thrale was the type of new industrialist, justly anticipating for his family the status of gentlefolk. Mrs Thrale, it cannot be doubted, was a product of decaying elegance . . .

It can be argued that Streatham Place was a setting where three different social milieux would intersect on the basis of a practically sufficient overlap of relevances and like-mindedness concerning the 'good life'. However, this association of the like-minded did not extend towards members of the local community. While they might have encountered each other on a Sunday morning at St Leonard's church, a place that Johnson was fond of allegedly, the members of the 'Streatham Place coterie' and the members of Streatham's local community did not actually participate in each other's social words. If we can rely on what is mentioned, or rather not mentioned, in the sources accessible, then Streatham Place existed quite detached from the rest of Streatham and its people. To take up Vulliamy's point here, the 'obliteration of strata' at Streatham Place did not include the local stratum. In other words, what we find is different social milieux coexisting in the same locality and yet not interfering with each other. Each milieu moves in what could be described as its respective 'sociosphere', or, in other words, a 'field of concern and relevance' that reflects the pattern of its social activities, evolving around a 'network of social relations of very different intensity', and 'spanning widely different territorial extents' (Albrow 1997: 51).

This perspective can be further illustrated by the fact that when Henry Thrale died in 1781, the social life around Streatham Place declined. The 'Literary Club' refocused on London after the death of its Streatham patron. Streatham Village, on the other hand, continued its tranquil existence as a 'pleasant rural retreat' (Gower 1990: 19). Even Johnson, who is said to have spoken of Streatham affectionately as 'my home' (ibid.: 15), finally left Streatham in 1782. Only his dedication on the burial plates of the Thrales in St Leonard's church still gives symbolic evidence of the ongoing flirtation between Johnson and Streatham Place.

But in what sense was it his 'home'? It is safe to assume that Dr Johnson's affection for Streatham was related to the hospitable atmosphere at Streatham Place, to the inspiring conversations at the Thrales' opulent dinner table, and to the walks with friends in the adjoining Surrey countryside. It was predominantly the option to associate with the like-minded that made Johnson feel at home in Streatham Place, rather then a deep and lasting involvement with the intimacies of the local community. Johnson and the others' attachment to Streatham was related to the spirit of Streatham Place rather than to the actual place Streatham as defined by its 'locale' and its local social relations. It is certainly no overstatement to suggest that the constellation of Dr Johnson's Streatham is an exemplification of the early beginnings of the delinking of milieu and locale.

It makes sense, then, to understand this Streatham constellation in the context of Pahl's argument concerning the difference between the social 'networks' of the privileged and the 'communities of common deprivation'. Although 'physically *in*' the locality of Streatham and its local community, the milieu of Johnson and friends is 'not socially *of* it' (Pahl 1973: 103f.). The constellation prevalent in Dr Johnson's

Streatham is a first indication of the stratification of different milieux in one locality, a stratification that relies on the exclusionary tendencies of access to new means of transport and communication. As the next section will further illustrate, by 1781 travel between Streatham and London, though technically improved, was still a privilege beyond the everyday lives of most people. So, on the one hand the Streatham of that time contained the *milieu of the privileged*, able to extend and withdraw its field of action and experience, and based on like-mindedness rather than locale. On the other hand, there is the *milieu of the locally confined*, deprived of the social luxury to choose appropriate local attachment according to shifting social networks of the like-minded.

A hundred years later this stratification of milieux linked to different sociospheres repeated itself with another famous name at its centre. From 1880 until his death Henry Tate (1819–99) took residence at 'Park Hill' mansion adjoining Streatham Common. Having made a fortune in the sugar trade, he started his career as a patron of British art after taking residence in Streatham. In fact, what was later to become the National Gallery of British Art, today known as the Tate Gallery, was developed and maintained at Park Hill. It was only in 1897, two years before Tate's death, that the gallery, 'adorned with the best works of contemporary masters', was moved to its present site at London's Millbank. So once again, Tate and his gallery added to Streatham's reputation amongst London's fashionable and well-off. At least once a year Streatham was at the centre of London's art scene, as 'every year, just before the opening of the academy exhibition, he [Tate] gave a dinner of the proportions of a banquet to the leading artists in his house' (*Dictionary of National Biography* 1901: 378f.; Gower n.d.: 9). However, as with the discussion groups at the Thrales' dinner table, the gallery at Park Hill was not open to the local public. Today the Tate Public Library, opened in Streatham High Road in 1890, is a reminder of Tate's residence in Streatham. But both the dedication – 'a gift to the people of Streatham from Sir Henry Tate' – and the fact that similar gifts were made to Brixton and Liverpool, suggest a rather detached attitude towards Streatham's local community (Gower 1990: 24; Weinreb and Hibbert 1983: 834).

In George Pratt (1820–98), a contemporary of Tate's, a property developer and founder of the aforementioned department store of the same name, we might see a contrasting figure with regard to involvement in and with the local community. His career is intrinsically interwoven with the development of Streatham during the last half of the nineteenth century. Starting his career as an apprentice with another local merchant, he remained dedicated to the locality at the peak of his business career when Pratt's became a household name in South London (Gower 1990: 26). Myers (1949: 130f.) reports in this regard that

Pratt's Department store made Streatham High Road one of the most attractive in South London, and joined by the 'Streatham Hill Theatre',

105

the 'Locarno Dance Hall' and the cinemas, the High Road could in a justified way be described as one of the busiest main streets in London.

To paraphrase Pahl again, Pratt and his commercial empire were clearly 'of the community', not just 'in it'. From this perspective, then, it is hardly surprising that the closure of Pratt's in 1990 was so symbolically (over)charged, and that the future of Streatham as a place within the urban landscape of the world city should evolve around that department store.

Pratt's role in Streatham's development was more complex than that though. His overall commercial success was linked to his involvement in the suburbanization of Streatham from 1850 onwards. So while on the one hand he profited from Streatham becoming part and parcel of one of the largest suburban areas in the world, the impact of his department store was to guarantee Streatham, at least initially, a reputation as a distinct and prosperous part of South London's suburbia.

The metropolis absorbs Streatham

The previous section showed that Streatham has always attracted those fragments of the London society that could afford to commute. As exemplified in Dr Johnson's Streatham, the privileged access to means of transport and communication by the old gentry and the new industrial middle classes initially led to a stratification of social milieux. Social stratification expressed itself in spatial mobility and relative independence from particular places, meaning the luxury to develop attachment to places of choice. This began to change during the nineteenth century with regular and public means of metropolitan transport becoming accessible to increasingly everyone. As Pahl (1973: 104) points out in this respect, the disruption of local communities really only starts when the less privileged can follow the mobility standards set by the well-to-do.

For Streatham this process, which saw it turn from a pleasant rural retreat into an Inner London suburb, started with the introduction of a regular coach service in the early 1800s. The service would go eight times a day in both directions, the journey taking an hour. Offering 'easy access to Westminster and the City', it was clearly geared towards the privileged commuter. This introduction of a regular transport link between London and Streatham did not as yet affect Streatham's rural tranquillity, but it indicated changes to come that would turn a rural town of 2,729 people in 1811 into a town of 20,000 by 1881, and then into one of its semi-detached suburbs of 70,000 inhabitants by the turn of the century (Weinreb and Hibbert 1983: 834). Or as Gower (1990: 21) puts it in a more life-world related perspective:

Of those in Streatham who celebrated the accession in 1837 of Queen Victoria to the throne, few would have believed that during her reign of

63 years their village would suffer a transformation hitherto undreamed of. For progressively throughout her reign the rapid expansion of London was to engulf many of the surrounding villages, and turning them into suburbs and destroying for ever much of their traditional identity.

Once again, the changes taking place in Streatham were a reflection of changes taking place in London, which in turn were determined by London's changing role in the world. The metropolis turning world city underwent drastic changes during the Victorian age. The City of London turned into the 'bank of the whole world', leading to an almost complete restructuring of the City between 1831 and 1881. Residential buildings were largely replaced by office blocks. The City was no longer a place of residence, but a centre from which international finance and banking, international companies, and the British Empire were managed. This pushed many of the former residents, mainly poor immigrants, into the outskirts (King 1991: 78; Seaman 1973: 9). At the same time, in the course of industrialization, the London area faced a rapid growth of its population, from 1 million in 1801, and 2.6 million in 1851, to 6.5 Mio in 1901 (King 1991: 79). Both factors led to what later became known as the 'suburbanization' of London, or more precisely the London 'south of the river', which during that process was to become 'the largest urban concentration which the world had ever seen (Brandon 1977: 111; Thorns 1972: 36).

As Thorns rightly suggests, suburbs south of the river Thames are not a recent phenomenon, but are part of London's history since the opening of London Bridge in 1209. What is of crucial interest here, despite the continuity in South London's suburban history, is the transition of the 'country house suburb' into a 'mass suburb' via cheaper and more frequent means of transport. The start was made with coaches and buses, with more than 100 routes already criss-crossing South London in 1834. From 1850 onwards the developing network of railway lines extended London's catchment area further towards the Surrey countryside. And the Cheap Transit Act of 1883, which led to the further reduction of transport fares, meant that metropolitan transport was no longer a privilege of the few (Thorns 1972: 36ff.; cf. Borer 1977: 179.).

Brandon (1977) suggests the following logic for the unfolding of the suburbanization that was to make Streatham a part of the metropolis. The first stage was marked by the opening of Westminster Bridge in 1750, linking Central London directly with the approach roads leading to Dover, Brighton and Portsmouth.

These were the great coaching arteries along which houses first slowly crept in the 1780s and then were to develop completely the ancient tightly clustered and well filled villages of Camberwell, Peckham, Stockwell, Streatham and Clapham.

(Brandon 1977: 110)

Brought into reasonable distance of the London society, rural retreats like Streatham trigger the 'new fashion to leave London at the weekend' (ibid.: 92). The second stage of suburbanization, which began around 1800 was then characterized by the steady enclosure of the common land between those towns and villages. The built-up area of London reached Kennington and the Oval. It had not yet reached Streatham (ibid.: 110). At this time a journey to Streatham was still going 'through a lot of common land where travellers could let their cattle graze' (Sexby's History of Streatham Common 1989).

However, the 1830s then saw a 'full-scale invasion' progressing. The built-up area of London rapidly extended south, while at the same time towns like Peckham, Dulwich, Clapham and Streatham 'were spreading amoebae-like towards each other. By 1868 they were solidly part of London' (Brandon 1977: 110). Another 30 years went by, and the aforementioned towns became part of what was later to be named 'Inner London', while the built-up area of London had already moved another six miles further into Surrey (ibid.: 111).

As far as Streatham in particular is concerned, the transition into a mass suburb was brought about by the railway. The Crystal Palace and West End Railway opened its first station in 1856. Two more lines went through Streatham by the 1860s, opening stations in 1862 and 1868 respectively. Streatham's connection to South London's transport network was then complemented by bus links, and also tram lines (Gower 1990: 22ff.; Jackson 1973: 40).

Streatham's incorporation into the metropolis, and the loss of what was perceived as autonomy from the big neighbour, finally manifested itself in administrative changes. The local administration by the Parish Vestry could not cope with a population of 70,000. Following the London Government Act of 1899, it was consequently replaced by the London Borough of Wandsworth and its council. This borough also included Tooting and other adjoining neighbourhoods (Bromhead 1936: 11; Gower 1990: 25). What looked quite logical and reasonable in terms of administration and infrastructure, at least from the observer's point of view, was experienced in quite a different way by those actually involved in these radical changes.

Struggle for local identity – the disruption of a local milieu

Time–space compression through improved means of (public) transport challenges the relative autonomy of place. As Harvey (1996: 279) aptly remarks, 'when transport costs were high and communication difficult, places were protected from competition by the friction of distance'. The integration of Streatham into the world city illustrates this assumption. It was no longer a relatively autonomous place, easily manageable in its community affairs, and clearly geographically marked by

108

neighbouring towns and the Surrey countryside. Streatham no longer satisfied the Aristotelian ideal of an urban unity, able to be 'taken in at a single view'. Streatham now had to maintain an identity amidst the somewhat anonymous landscape of South London's suburbia. The town was now certainly an 'open' place in Harvey's sense (1996: 310), exposed to the socio-ecological processes shaping the world city.

The accessible sources of local history suggest that the local residents experienced the suburbanization of 'their' Streatham as a disruption of a local milieu, based on 'the constancy of the surrounding social and material environment of action' (Giddens 1994: 92). The coming of the railway, for instance, was perceived as a destruction of the local setting, as two railway lines literally criss-crossed Tooting Common. Bromhead (1936: 13) records the local resentment when he states: 'The convenience of the public is, I suppose, the excuse for the permanent disfigurement of our lovely Tooting Common which these lines have made'.

Moreover, the coming of the railway took away some of the reputation of Streatham as a rural retreat. Londoners seeking the pleasures of the countryside would now have to pass Streatham and go six miles further south. But more importantly, metropolitan time–space compression made places accessible to the general public that were formerly only accessible by the privileged. As Arnold (1886: 104) observes, the metropolitan railway network 'brought distant places of interest within the same time and purses of "cheap trippers"', for example, in 1854, when Crystal Palace re-opened a few miles south of Streatham in Sydenham, where it was to become a permanent leisure and exhibition centre, attracting 2 million visitors a year (Reeves 1986: 23, 31). It seems to be more than a coincidence that the Streatham Wells, and the Rookery Tea Gardens attached to them, had to be closed by 1860, while they had still been attracting many weekend trippers in the first half of that century (Weinreb and Hibbert 1983: 835).

Equally drastic were the changes in the built environment. In order to cope with the influx of newcomers, a number of residential estates had to be built. In turn, a lot of familiar old mansions, reminiscent of Streatham's past as a fashionable rural retreat, had to be demolished. This also meant further adding to the semi-detached houses around the commons, furthering the suburban flair of the place (Gower 1990: 23, 27f.; Weinreb and Hibbert 1983: 834). The locals frankly felt overwhelmed by this massive development of new properties, nicely captured in Arnold's (1886: 216) slightly cynical statement that 'a few years ago a new house was a thing to be talked about'.

Though much later, of major symbolic significance for the changes taking place in Streatham as an integral part of the global city, was the opening of a McDonald's branch in Streatham High Road on 1 August 1979. As the recent debate around the opening of a McDonald's outlet in Hampstead, one of the last London villages, seems to suggest, resistance against global fast-food outlets is a recurrent pattern. Not surprisingly, as it makes the openness or the phantasmagoric character of the place obvious to everyone, no matter how much the local discourse emphasizes images of local autonomy. Moreover, in the case of Streatham, the opening of the McDonald's branch was much more symbolically charged. The McDonald's branch occupied a building that was formerly used by Pratt's department store. For the older generation in particular, the symbolic evidence of the alleged 'decline of Streatham' could not be more striking. With the closure of Pratt's in 1990, the image of lost autonomy and helplessness against decisions taken elsewhere, became complete.

However, going back in time again, given that local identity was to become increasingly a discursively produced aestheticized image, the major challenge to Streatham's local community was the influx of 'newcomers' who did not necessarily share with the locals a knowledge of the Streatham of the past, and who might not feel obliged to join the discursive construction of local identity within the urban flux of the metropolis. Bromhead (1936: 3) illustrates this rupture in the symbolic discourse:

> The Streatham of the secluded houses, of fields and lanes that some of us knew and loved has passed; but the past still holds us in willing service for Streatham and Streatham people. For those many who have since come amongst us, an interest in a Streatham that is to all appearance fast losing its identity – one might almost say its soul – in a Greater London, must be awakened.

The transformation that turned Streatham into an Inner London suburb turned out to be of lasting consequence. Streatham was not simply to reposition itself within an altered but stable geography of space. Incorporated into the world city it was now part of the socio-ecological processes that shape the metropolitan urban landscape. This meant that while the reality and the reputation of Streatham the 'rural retreat' was a thing of the past, metropolitan life offered other options to develop an alternative image. During the period before and after the Second World War, Streatham was able to develop the reputation of being the 'entertainment centre of South London' (Gower 1990: 28; Weinreb and Hibbert 1983: 834). With its theatre, dance hall, ice rink, and several cinemas, it was the place to be for regional entertainment and leisure. Here Streatham clearly benefited from the integration into the metropolis, as these facilities relied on a wider area than just the local catchment to run profitably. However, that from the 1960s onwards Streatham lost part of that image due to the closure of theatre and dance hall, while other South London areas such as Wimbledon and Clapham gained reputations as trendy places, only goes to show the competitiveness within the metropolitan 'economy of signs and space'. During the 1970s and the 1980s Streatham cultivated its reputation as a centre of leisure and good shopping but, coming back to where we started this chapter, following the closure of Pratt's in 1990, even that image is now under threat.

What becomes obvious to the participant observer in Streatham after a while is that an image of Streatham's past is passed on to each following generation of 'committed' residents. The Streatham of the 'secluded houses, fields and lanes' is followed by the 'West End of South London', in turn followed by the Streatham of 'shopping at Pratt's'. Each is considered to be a decline in comparison to the previous one, while the struggle for local identity in the 1990s is simply against the latent danger of being associated with the negative image of 'just another part of South London suburbia'. The continuity between the Streathams of the past and today's Streatham in the local discourse is maintained mainly through the two resident associations. The arrangement of these associations illustrates Pahl's (1973) argument concerning 'formal voluntary associations' and the discursive role of the educated middle classes within them. On the one hand, they integrate numerous members of the older generation, who can still recall the fashionable Streatham of the theatre and the dance hall. And on the other hand, they are dominated by people in middle-class careers and life styles, who are able to reformulate the local narratives of the older generation into a rational argument for the need of future developments for Streatham. While the former talk about the Streatham of 'those days', the latter simply utilize these narratives of a bygone Streatham for a symbolic image of Streatham that helps it to get 'back on the map of London', to quote the phrase mostly used in the meetings of both associations. This seems to be a quite complementary arrangement, with each side balancing the other.

As pointed out earlier, the continuity between the Streatham of the past and today's Streatham is mainly anchored in the boarded-up ruin of what used to be Pratt's department store. What seems surprising at first to the participant observer, namely the rallying of a locality around the ruin of a department store, makes sense when taking the symbolic significance of the site into account. The story of George Pratt and his department store is intrinsically intertwined with the commercial success and symbolic attraction of Streatham's High Road. Its closure is a symbolic marker for the 'decline' of Streatham as an area, a symbol of struggle for local identity in the present, and a focus of different people's hopes for a Streatham of future developments. Harvey (1993: 272), following Simmel, points out that ruins are particularly useful for the social construction of local identity, for they are 'places where the past with its destinies and transformations has been gathered into this instant of an aesthetically perceptible present'.

Moreover, this interplay of past, present and future in one building highlights again the constitutive and generative element in people's milieux. Different milieux will have a different perspective on the same material site. And indeed, what for the 'committed' residents of Streatham is still affectionately known as Pratt's, is for others just another demolished building along Streatham High Road. What this particular example also illustrates is that the milieu transcends the immediate locale, not only in space, but in time. It goes beyond the actually 'perceptible present' when

111

people refer the interested newcomer to the site of the former department store by saying: 'You *see*, this is where Pratt's used to be'. Pratt's, although (or maybe because) it is physically no longer there, is still something that practically affects many people's everyday lives in Streatham, and, consequently, is still part of their practical milieu (de Certeau 1988: 108).

Scheler (1973: 147) points out in this regard that '*tradition*', namely 'history as living and effective in us', should be understood as the 'temporal extension of milieu'. In this regard, then, Pratt's can be seen as functioning as an anchor of tradition. It symbolizes the identity of a locality that has undergone irreversible trans-formations over the past decades. As a local site that symbolizes 'those good old days', it helps people to live in a *milieu of local tradition*. This kind of living in the past in order to find comportment towards a world in irreversible transition, is quite explicit in the following statement from a leading voice in the discourse that maintains and utilizes images of a bygone Streatham:

> That's pride, that's deep rooted pride. The people of Streatham still feel that they want to hang on to that rather nice way of life that they have known . . . from about 1949–50 into the 1960s, even up to about 1974–75, when they felt as though they were a cut above of the other parts of the Borough of Lambeth. After all, they had a very big department store which was very well known, and the High Road contained a lot of major retailers. It was a good place to live, a good place to work, good place to do shopping, and a sought after area in which to live.
>
> Now the situation has changed. But still, in the minds of *a lot of people, they still live in those 'good old days'*, perhaps twenty years ago. They still haven't changed the way they think about Streatham. I'm pleased that they don't. Because I don't think it would be right to allow Streatham to go down. It needs to be kept at a certain level.
>
> But of course, the people who live here are vastly different, . . . there are lots of flats now, a lot of movement in terms of people moving in and out of properties, and crime has increased in Streatham in the last five or six years quite dramatically. And, it's actually very worrying, the nature of the High Road has changed.
>
> (C. Barnett,[2] Committee member of the 'Streatham Association';
>
> my emphasis)

What is being acknowledged in this account is that the discourse on Streatham's local identity is complex and not shared by everyone living in Streatham. People live in different milieux despite coexisting in the same locality. The freelance consultant is likely to be more interested in the history of his area than is his daughter who has decided to live in a squat in Brixton. In addition, images of the past are a contested symbolic site. What is interesting to observe in this regard is a 'metropolitan turn' in Streatham's images of local identity. Whereas formerly the emphasis was on the locality's reputation as a place distinct from the neighbouring metropolis, there is now a certain tendency to claim London symbols and London images for the locality. The emphasis on Dr Johnson's engagement with Streatham seems to be a rather justified and plausible claim in this regard. Only recently 'discovered' by the local discourse, but more important, is the alleged presence of Sherlock Holmes in Streatham. The author Arthur Conan Doyle resided in Norbury, an area in the neighbourhood of Streatham from 1891 till 1894. As a passionate user of the bicycle, he is supposed to 'have peddled through Streatham on many occasions as the High Road is the most direct route to Norbury'. For whatever reason, he located one of his stories in Streatham, which enabled the 'Local History Reprints' (Gower 1993) to claim:

> Amongst the many interesting people who visited Streatham during Victorian times none is perhaps more interesting than Sherlock Holmes, the great consulting detective who, during the space of 24 hours, came to Streatham on no less than three occasions in order to solve the mystery of 'The Adventure of the Beryl Coronet'.

The young generations, however, are more likely to refer to Naomi Campbell, the Streatham-born super model, who provides living proof for the claim that someone from Streatham can make it in London and the world. The super model's mother still lives in Streatham as a local celebrity and occasionally attends the opening festivities of a new local entertainment venue. These are ways of claiming a symbolic reputation for the area that are clearly more outward looking than the revelling in local images of a bygone Streatham.

Referring back to the starting point of this chapter, it appears, then, that contemplation and historic review must have their place in metropolitan life. Discursively constructed loyalty to a place helps generate situatedness in the transitory and ephemeral urban landscape of global city. It can also generate a degree of local solidarity between otherwise coexisting milieux that is sufficient to lobby for local needs. Conversely, romanticizing and nostalgic indulgence in the local past potentially discourages engagement with the problems of present life in the locality, as an imperfect present sits uncomfortably against an idolized past. Nostalgic localism at best alienates those not prepared to join the discourse of local history, at worst it can serve as a means to put social pressure on the outsider.

There is no clear-cut boundary between the historic grounding of 'place' within urban space and nostalgic localism. Changing spatial and social constellations will require different degrees of engagement with local history, as seen in the case of Streatham and the changing images of the past. What comes to mind here is

Nietzsche's argument concerning 'the use and abuse of history' (1874). According to Nietzsche, both 'the historical and the unhistorical are equally necessary to the health of an individual, a community, and a system of culture' (1979: 8). In general, the concern with history is commendable, as long as studying history serves practical life. What amounts to the 'right' measure of history, according to Nietzsche (ibid.: 7), depends on how much it feeds the 'plastic power' of an individual or a community, which means,

> the power of specifically growing out of one's self, of making the past and the strange one body with the near and the present, of healing wounds, replacing what is lost, repairing broken molds.

Seen from this perspective, then, cultivating images of the local past does not immediately equate simplistic parochialism. Instead these images help to constitute the symbolic 'horizon' that serves both stability and adaptability within a transitory and ephemeral world (Nietzsche 1979: 7). Harvey (1993: 288) argues in a similar vein when discussing the function of 'images of permanence' within contemporary society. He points out that:

> [they] require considerable sophistication, because the continuity and stability of the image have to be retained while stressing the adaptability, flexibility, and dynamism of whoever or whatever is being imagined.

The discursive construct of the Streatham of the past can be seen as one such possible cluster of images that provides comportment for the daily engagement with the hustle and bustle of London life. It depends on the actual disposition of the individual's own milieu as to whether this symbolic reassurance leads to positive engagement with what metropolitan life has to offer, or to backwards-looking escapism.

7

EXTENDED MILIEU AND 'SOFT CITY' – GENERATING SYMBOLIC SPACE

> Cities, unlike villages and small towns are plastic by nature. We mould them in our images: they, in their turn, shape us by the resistance they offer when we try to impose our own personal form on them.
>
> (Raban 1990: 10)

The process of time–space compression has, from the outset, double-faced implications for people's everyday lives. While the world seems to shrink in terms of technical access to distant places, people, and cultures, the life-world of each of us extends, both in scope and content. While on the one hand we can observe the extension of people's milieux across distance, there is, on the other hand, the penetration of distant worlds into the actual 'here and now'.

The latter process was earlier described as *microglobalization*, referring to the compression of global sociocultural flows into distinguished sociocultural environments, such as the world city. The globalized world city itself can now be regarded as a pastiche of different locales, with a potpourri of global culture evolving around them. Subsequently, the milieux of those living the globalized world city will potentially stretch across a patchwork of different London localities and the experiences of fragmented sociocultural worlds.

The notion of an extended milieu discussed so far in this book, has stressed the enlargement of people's 'symbolic territory of the self' *beyond* the locality of London. It shall now be argued that a similar process of *extension en miniature* takes place for the milieu *within* the globalized world city. Referring back to Raban's argument, this chapter pursues the claim that generating a 'Soft City' in a place like London means in fact generating the symbolic space of an extended milieu.

'Soft city' and 'concept city'

Ever since Aristotle's description of the Greek *polis*, there has been the ideal notion of the city as an urban unity that 'can be taken in at a single view'. Following Mumford's reading of Aristotle (1991: 216), there are basically two reasons, one aesthetic and the other political, for deliberately limiting the size of the city. First, the overall view one gained when looking down from the height of the Acropolis meant that the city took on the familiarity of a single person, so to know its form and character was an essential precondition for successfully pursuing the 'good life' that the polis was supposed to convey. Relatedly, the political concept of the polis demanded from its citizens the ability to command and to judge for the whole on the basis of personal acquaintance with each other and with the environment they shared. This principle of urban social order was clearly demanding limitation in size and numbers.

The city in history has long since passed the ideal size of the polis, and yet, the notion of an ideal urban unity persists, both as a means to generate familiarity in the urban life-world on the one hand, and as a means to guarantee administrative control on the other. Nowhere more so than in the contemporary metropolis does the ambiguity come to the fore that is inherent in the small-scale urban unity as an organizing principle of both the life-world and the administrative system.

Harvey (1993: 5ff., 66ff., 82f.), for example, in his analysis of the urban environment as a mirror of the postmodern condition, places the strongly defined and visible urban unity that is achieved by modernist urban planning and architecture, against the loosely defined and largely invisible urban unity that is generated by the highly differentiated practical discourses of a heterogenous and shifting urban population. The modern city exemplifies the modernist attempt to control social life and human behaviour via the rationalization of spatial patterns, most radically applied in Hausmann's slum clearance of Paris in the 1860s. But urban development since, from the Paris Commune of 1871 to contemporary urban social movements, has shown that metropolitan life is far too complex and creative a phenomenon to ever be disciplined by a spatial order. To paraphrase Harvey's argument here, urban practices have shown the 'awkward habit' of breaking through the 'tyranny of the straight line' and generate their own urban orders (ibid.: 204; see also Kiwitz 1986: 142f.).

A similar argument is brought forward by de Certeau (1988: 93ff.), who considers the 'concept city' as the product of a utopian modernist urban discourse, which attempted to integrate the multiplicity of 'urban practices' into an 'urbanistic system'. The 'concept city', according to de Certeau, embodied 'simultaneously the machinery and the hero of modernity', allowing for urban space to become the playing field of a totalizing rationality and technology. He goes on to conclude that the 'urbanistic project' could never have succeeded, for it failed to discipline the spontaneity of urban practice.

De Certeau (1988: 94ff.) goes further in his argument than Harvey insofar as he does not only state the end of the totalizing 'concept city', but radically stresses the priority of urban practice over the built environment. His interest is focused on the 'microbe-like singular and plural practices that an urbanistic system was supposed to administer or suppress, but which have outlived its decay'. According to de Certeau it is those everyday practices that 'reappropriate' the 'disciplinary space' generated by the built environment, and which in turn are the cataclysts of 'lived space'. De Certeau describes this dimension of the city, which is generated by the plethora of urban practices inscribing themselves into the fixed forms of the built environment as the 'migrational or metaphorical city' (ibid.: 93).

It is this phenomenon of a reappropriation and realization of urban space through the multiplicity of individual and social practices of urban everyday life, that Raban (1990: 10) has imaginatively described as the 'Soft City':

> For better or worse, it [the city] invites you to remake it, to consolidate it into a shape you can live in . . . In this sense, it seems to me that living in cities is an art, and we need the vocabulary of art, of style, to describe the peculiar relationship between man and material that exists in the continual creative play of urban living. The city as we imagine it, the soft city of illusion, myth, aspiration, nightmare, is as real, maybe more real than the hard city one can locate on maps in statistics, in monographs on urban sociology and demography and architecture.

Symbolic space and the concept of the city

It would be too simple an exercise just to confront 'concept city' and the 'soft city', and to equate them with the opposition of an imposed systemic order on the one hand, and a freely generated order of the life-world on the other hand. This is what de Certeau (1988: 96) seems to express rather drastically when he links the inherent ambiguity within the idea of an urban unity to the conflict between 'disciplinary space' versus 'lived space'. We have to recall here that Aristotle's idea of the city as a single unity did not just refer to the imposition of a politico-administrative order, but was also inspired by the aesthetic and life-world related motive to become familiar with the city in its entirety.

This point is implicitly taken up in de Certeau's analysis of the 'practiced city' (1988: 91f.). Describing the bird's eye view of Manhattan from the 110th floor of the World Trade Centre, de Certeau poses the question, What actually is 'this pleasure of seeing the whole', of looking down on the city as a unity? His answer is that this overview seems to bring the city into our 'grasp' by seemingly allowing us to escape the city's grasp, in which we are normally caught in the normal course of our everyday lives. However, the price we pay for 'grasping' the city in an overview, according to de Certeau, is distance. Distance from the many interwoven practices that make up the human fabric of the city. Does this mean, then, that the pleasure we gain from seeing the city as a coherent geographic unit is irrelevant for the 'practiced city'?

There is a certain comfort in having, at least for the moment, the abstract geographical identity of the city laid out in front of us when, otherwise, our attempts to gain a comprehensive picture of the metropolis as a lived experience are confined to the linking together of specific fragments of the city in our makeshift mental maps. Taking in the city as a whole from a distance provides a different kind of acquaintance with the city compared to the fragmented familiarity we have to maintain in the complex and shifting landscape of the 'practiced city'. It is, then, perhaps not surprising that especially the 'newcomer', trying to find initial comportment towards the overwhelming rhythm of the big city, should feel inclined to look for the reassuring perspective on the 'whole'. That will not always be possible via the bird's eye view from the World Trade Centre. Barbara, the American expatriate who recently moved from Manila to London explains, referring to her initial period of settling into London:

> I'm just beginning to get some, some perspective. It's pretty overwhelming to get a perspective of the city, since I haven't really been able to see how it's laid out. I'm only beginning to get some idea from the subways, but then you are underground, you can't really tell what the scene is up above, and so when I come up from the underground at a different place, I can't really get the relationship between the other places I've been.

Referring back to de Certeau's argument, what this interview sequence indicates is the life-world related motive to develop a *concept of the city* that helps to integrate and stabilize a 'soft city' in the making. This is not to be confused with the rationalized unity of the 'concept city' of urban planning and administration. Barbara's desire to visualize the 'whole' of London and 'how it's laid out' does not derive from what de Certau describes as the god-like 'lust to be a viewpoint and nothing more', detached from all urban practice (1988: 92). Instead, Barbara's interest in a perspective of the whole derives from the more mundane need to re-establish her milieu, to conduct routines and pursue plans in an (as yet) unfamiliar environment. This is clearly expressed in her summarizing statement concerning the period of settling into London: 'I am trying to find *my way* in the big city' (my emphasis). The desire to visualize the complex environment of the big city can then be understood as one moment of that attempt to regain a practical milieu.

To fully appreciate this interpretation of Barbara's desire to get a grasp on the layout of the whole of London, we have to recall some of Scheler's and Schütz's ideas concerning the milieu as a form of personal orientation in the world. From Scheler

(1973: 141ff.) we have to reconvene that milieu in the first instance refers to the individual's value-related general disposition towards the world, a mind-set of sorts that works simultaneously as a guide and filter in the engagement with our surroundings. Thus, the milieu directs us towards 'points of departure of possible acting' within an (as yet) unfamiliar environment. But, as Scheler stresses, it is only '*in* the execution of acting' that the environment which carries our 'milieu-structure' is actually realized as a meaningful configuration. This interplay between milieu and environment, or 'milieu-structure' and 'practical milieu' in Scheler's terms, is an ongoing process, open for ruptures and distortion. However, under normal circumstances milieu and surrounding environment are assumed to 'coincide' to a degree that is practically sufficient (ibid.: 134).

Schütz (1966: 122f.) on the other hand explains the problem of (re-)orientation in a new environment by drawing on his concept of 'relevancy'. To re-establish a milieu implies to bring an (as yet) unknown environment into the perspective of the individual's 'biographical situation'. Orientation in the context of Schütz's argument, then, means to identify

> which elements of both the ontological structure of the pre-given world and the actual stock of knowledge are *relevant* for the individual to define his situation thinkingly, actingly, emotionally, to find his way in it, and to come to terms with it.

Schütz also stresses that the process of (re-)orientation is an ongoing one. Changes in our surroundings, as much as changes in the individual's 'biographical situation' (life-plans, stock of knowledge, habitualities, and so on), imply that our practical milieu will be restructured 'in both respects – as to space and time – in strata of major and minor relevance' (1967: 212f., 227).

Looking at Barbara's remarks against this theoretical background, we can argue that her project of 'trying to find *my* way in the big city' is not a plain geographical search for direction, but a symbolic expression of her attempt to realize 'my milieu' in London's everyday life. We should not be surprised to find both scientific (Schütz) and everyday life (Barbara) discourse using a spatial metaphor – 'to find your way' – in order to express the process of realizing intrinsic, that is motivational and thematical, relevances of action. For a crucial aspect in generating a 'practical milieu' is the spatialization of intrinsic relevances.

In Barbara's particular case this means engaging with different relevant locales, for example, her flat in Streatham, her office in Vauxhall, St Anne and St Agnes Church in the City, the British Museum, the Wimbledon Tennis Courts, and places of residence of Filipino friends. She will have to engage in social practices that are not locally fixed, but progress across territories, say, participating in the social activities of London's Filipino community, enjoying London's museum and theatre

119

land, commuting between 'north and south of the river', and so on. The real challenge for the newcomer is to connect these places, people, and social engagements across distance, not only in a technically convenient way, by simply relying on systemic linkages such as the underground network, but instead to provide these linkages with a biographical index. This desire to emancipate 'her' London from the London of the underground network or the 'A–Z' map is expressed in remarks such as wanting to see 'what the scene is up above', to '*really* get the relationship between the places I've been', and once again, 'finding *my* way in the big city'.

It follows from this that the situatedness of a practical milieu is not solely a function of the individual's milieu-structure 'fitting' the environment at hand in terms of content. Spatial situatedness is equally important. In other words, the embeddedness of the practical milieu in a consistent configuration of meaning, given, amongst other things, by the direction of life-plans and the continuity of daily routines, has to be complemented by the individual's competence in handling the milieu as a manageable and coherent spatial configuration.

Consequently, it would appear that a crisis of (dis)orientation in the newcomer's milieu can potentially have two dimensions, a content-related and a symbolic one. The former reflects on the failure to link crucial relevances of action to the relevant aspects of the pre-given environment. This would be the case if Barbara, for instance, had failed to realize her 'Filipino London', thus truncating her London milieu by a crucial dimension. It is perhaps one of the distinguished features of life in the globalized world city that such a rapture would be unlikely to persist, given the social and cultural variety which a microglobalized environment offers.

Second, however, symbolic disorientation can arise when places and practices that constitute the individual's milieu cannot be integrated into a coherent representation of their relevant spatialities. Following Cassirer (1957: 244f.) we could say that symbolic disorientation implies the inability to realize the practical milieu as a 'symbolic space'. Looking at it from this perspective, Barbara's desire to see the 'whole' and 'how it's laid out', can then be understood as an indicator of a temporary crisis in her attempt to realize and generate the relevant symbolic space of her London milieu. As yet, by her own account, Barbara feels uncomfortable with exploring and 'flâneuring' in London, and consequently, she prefers to stick to the bits she already knows, strictly focusing her routines to and from home. At the same time she indicates that this might be a temporary problem, gradually disappearing as the process of settling in continues. Asked with regard to her interest in exploring London she says:

> No, not at this point anyway, because I just don't feel comfortable with going somewhere without being reassured how I get back home again. If I become a little more familiar, I'll be able to feel a little more free to do that.

The message in Barbara's remarks can be clarified further when looking at Cassirer's (1957: 243, 158) crucial differentiation between 'active space' and 'symbolic space' as two ideal types of spatial configurations. While the former is defined as 'mere field of action' (*Aktionsfeld*), the latter is best described, following Leibniz, as 'an order of possible coexistence' (*Ordnung des möglichen Beisammen*).

The two types of space are constituted by two different 'modes of orientation'. In 'active space' the individual's actions are orientated towards sensory impressions of the immediate environment. Familiarity in this context is based on detailed knowledge of specific situations and actions becoming 'fused with those situations'. Subsequently, 'stereotyped' actions become fixed in 'customary channels' (Barbara saying '. . . how I get back home again') and cannot be adapted to 'unaccustomed circumstances'. This clinging to detail of the immediate surroundings and routine, according to Cassirer, is due to an 'incapacity for schematization' (in Barbara's words '. . . I haven't really been able to really see how it's laid out') (Cassirer 1957: 171f., 243ff., 270ff.).

In contrast, orientation in 'symbolic space' implies the 'free survey of spatial determinations and relations' (Barbara: 'I can't really get the relationship between the other places I've been'), and the ability 'to compose locally separate factors into the unity of a simultaneous view' ('I'm just beginning to get some perspective'). Symbolic space then provides a scheme in which the departure point and direction of action 'can be freely chosen and shifted at will' ('I'll be able to feel a little more free to do that [flâneuring]'). Situatedness in symbolic space, then, means the ability to detach actions from the consistency of detail and to anticipate, play with, and realize the multiple options of the spatial environment relatively independent of the actual bodily location (ibid.: 152f., 158f., 243f.).

It is, moreover, important to note that, still following Cassirer's argument, generating symbolic space is based on gaining a 'unity of sight' (ibid.: 152f., 158ff.), though this unity does not derive from a bird's eye view from a skyscraper. Neither based on sensory data, as action space would be, nor on mere abstraction, as geographical space would be, symbolic space relies on the 'basic function of re-presentation'. It is essentially an 'intuitive space',[1] projected towards an imagined environment which takes in 'the real and the possible' and 'interchanges present and nonpresent'. Symbolic space derives from focusing on the relevant aspects, as much as disregarding non-relevant aspects, of the surrounding environment with regard to a certain task or plan. Consequently, the situatedness of action in symbolic space shifts from the continuity of particular circumstances towards the projection of an 'ideal plan, which anticipates the action as a whole and which assures its unity, cohesion, and continuity' (ibid.: 182, 243ff., 270ff.).

Cassirer metaphorically describes the generation of symbolic space out of action space as a continuous transition from 'sensory-physical grasping' (*Er-greifen*) to 'conceptual grasping' (*Begreifen*).[2] As Barbara's desire to 'see the whole' of London exemplifies, sensory-physical grasping in Cassirer's sense, implies that a phenomenon has to be possessed 'immediately with the senses'. Conceptual grasping, on the other hand, would imply detachment from 'sensory immediacy' and the gaining of a 'mediate grasp' or symbolic synopsis of the relevent aspects of the surroundings (1957: 181f., 243f.).

Barbara's desire to 'see the whole' of London indicates the symbolic space of a milieu in the making. She feels the need to bring the different action spaces in which she has begun to realize her practical milieu in London into an overall perspective or scheme. Her remarks convey what de Certeau (1988: 92) describes as desire 'to be lifted out of the city's grasp' (Barbara: 'It's pretty overwhelming . . .'), to gain appropriate distance, so that her actions, for the moment at least, are no longer 'possessed' by the big city's everyday life that follows an 'anonymous law'. At this stage of settling into London and regaining her milieu, Barbara clearly lacks the intuitive synopsis of her environs and the symbolic distance which Cassirer refers to as crucial for acting in symbolic space. The desire to see the geographic unity of the city laid out in front of her can only serve as a temporary substitution. For 'seeing' the city as single geographic unity would bring the overwhelming complexity of the city's everyday life into perspective, which is however, as de Certeau reminds us, a perspective of contemplation and disengagement with the practiced city.

Instead, the synopsis of symbolic space that Cassirere talks about is a unity of direction (1957: 182), an ideal plan of action of sorts, not detached, but also not completely absorbed by the immediate reality of our surroundings. It lends itself for a flexible concept of the city that displays the direction of the individual's milieu in its spatial dimension, not however an abstract spatial unity. It is a perspective of the practiced city rather than an overview of an abstract geographic unity of urban space.

Finally, with regard to the problematic relationships between urban unity and urban practice as raised by de Certeau, it can then be argued that, while the general contrasting of concept city and practiced city certainly holds, the simple counter-positioning of geographical space and lived space is rather problematic. In the light of the newcomer's concern with seeing the whole, and our interpretation of it via Cassirer's concept of symbolic space, it would appear that the lived space of the 'soft city' is not simply the outcome of spontaneous urban practices, but requires its own underlying concept of the city. Only that this unity of the whole, unlike the concept city drawn up by the urban planner, is an order of the life-world, representing the whole of the city insofar as practically relevant in the individual's milieux. Each milieu has its concept of the city. The London as 'seen' by a newcomer like Barbara will be different from, for example, that of a Londoner by birth, like Ulla, or the London of the global business man, like Harold. For each of them the symbolic synopsis of 'their' London reflects their engagement in different social, and that is always spatial, practices across the metropolis, in a relatively coherent and biographically meaningful configuration.

The generating of symbolic space in the context of a milieu is then not contra-dictory to the lived space of shifting urban practices. Instead, it can be seen as a condition for the emancipation of these practices from what de Certeau described as the 'grasp' of the autonomous rhythm of the city. To act in a milieu-related symbolic space, to perceive and manage the different London environments in relation to her 'biographical situation', is the basic condition on which Barbara will be able to turn the 'pretty overwhelming' experience of London into 'her' London.

The 'soft city' as extended milieu

The previous section outlined a shift in the spatial situatedness of a metropolitan newcomer's milieu. Barbara's remarks captured a milieu in the making at the turning point of integrating a multiplicity of rather unrelated active spaces into the more coherent configuration of symbolic space. In this process action is symbolic-ally disembedded from the sensory-physical make up of the immediate surroundings, and reembedded in an 'ideal plan' of action, enabling the pursuit of action relatively independent of alterations in the immediate environs. She has, in other words, uprooted her milieu from the restrictions of action space.

In practice this means, amongst other things, that Barbara will no longer be con-cerned with the technical detail of 'how I get back home again'. Instead of simply following the technically most convenient routes and clinging to signposting landmarks in her immediate sensory environment, she will develop 'her' own short cuts and detours where appropriate and in accordance with changes in her affective disposition or the task at hand. As she suggested herself, as time goes by she will 'feel a little more free' to explore and to be a flâneur, on her own or with her Filipino friends. In de Certeau's words, in re-establishing her milieu in London, Barbara will develop her 'individual modes of reappropriation' of urban space (1988: 96). Different places in the urban landscape of London will gain significance for her through memorable encounters, or because they are linked with the names of friends and colleagues.

Barbara's 'soft city' will not be entirely structured by eager anticipation of pleasant people and places, but also by fear and anxiety. Like Sarah, whose remark might illustrate this point, she will encounter her own 'no-go-areas', which effectively structure her soft city in a negative sort of way:

> For example, the Elephant and Castle I find a very depressing area, it's just the way it's designed and all the subways and that sort of thing . . . But I also find it unsafe because of the way it's designed. For that reason I try and avoid it while I'm using public transport, I avoid changing there for

example, and I would try to do my route a different way. So I don't let it worry me, I do work my way around it, to a certain extent, but that's fairly unconscious.

Barbara will eventually manage the symbolic space of her milieu in such a way that allows her to manipulate the spatial structure of the big city according to her own needs, just as Ira or Sarah manage to do. Ira, we recall, told us earlier on how she manages to find the places and meet the people that fit the 'vibes' she finds herself in. Rather than being in the 'grasp' of the city, Ira has the city more or less at her disposal. She knows where to sit and watch the American tourists leaving the obligatory musical performance, knows where to meet single men on a Friday evening, or where to find the pubs that have the right atmosphere for a bit of soul searching. Similarly, Sarah described how she effectively manages to separate work and leisure as if they were 'a world apart', despite their proximity in geographical space.

It is, however, quite interesting to note that Sarah, despite her confident way of managing 'her' London, admits to using London's famous 'A–Z' map quite frequently when she goes to one of her singing events, or when friends suggest meeting up somewhere in a restaurant that has just opened. This, it would seem, is somewhat disenchanting to Barbara's enthusiasm with regard to becoming familiar with the 'whole' of London. This little contrast may seem hardly worth noticing, yet it reveals something about the relationship between 'soft city' and 'urban unity' that is peculiar to the urban metropolis.

In a city like London, with its transitory urban landscape, the process of generating a 'soft city' is an ongoing project. Making new friends, moving house, a new hobby, gentrification, and many other happenings can trigger a change in the symbolic spatiality of the individual's 'soft city', forcing it to withdraw from known places, and encouraging it to conquer unknown territory. This in turn means that the (re)generation of symbolic space is a continuous process. Accordingly, falling back into what Cassierer typified as action space by, say, desperately looking for familiar landmarks that point the way back to tube station or car park, getting lost, and using the 'A–Z' as a readily available substitute for the bird's eye view, are somewhat embarrassing for 'the Londoner', and yet part of London life.

Relatedly, the sheer size of a place like London makes it impossible to be covered by someone's soft city. Someone's 'soft city' under these circumstances not only has the characteristics of a 'perspective', but moreover, takes the form of a pastiche or collage. This implies that, not only do we have to realize the relevant places and attach meaning to them, but we also have to ignore large chunks of the metropolitan landscape. In the big city, after an initial time of 'flâneuring' perhaps, we make choices, deliberate or unconscious, as to which parts are more relevant than others. For example, for Barbara, places like Southall, Ealing, Haringey are of no practical concern, and might as well be located in Aberdeen or Leeds or somewhere else in the world for that matter, for she has no intention of visiting these places during her two years in London. Moreover, unless one is prepared to be confined to the surroundings within walking distance, the milieu stretches across the urban landscape, taking in distant places and people. But it does not comprehensively cover the territory in between. In order to link these places in a manageable way, the metropolitan dweller has to make use of abstract systems like the underground network, which by nature deprives us of the immediate experience of the 'in between'. Differently from a village or town, living in a city of the size of London means accepting a ruptured and fragmented field of experience (Waldenfels 1985: 190ff., 204). Barbara clearly expressed unease about this when saying:

> I'm only beginning to get some idea [of the city] from the subways, but then you are underground, you can't really tell what the scene is up above, and so when I come up from the underground at a different place, I can't really get the relationship between the other places I've been.

Therefore, Raban's famous quote concerning the plastic nature of the 'soft city' could be slightly changed with regard to a city like London, insofar as it actually '*forces* you to remake it, to consolidate it into a shape you can live in' (Raban 1990: 9).

However, the more crucial point to be drawn from these observations is that the situatedness of a milieu in the big city cannot be based on the continuous experience of a coherent locale or local landscape. The metropolitan dweller lives in a fragmented field of experience and action that has to be 'consolidated' in symbolic space. Generating a milieu *within* the big city reveals in its partial organization the same general tendency as the one identified for the extended milieu that reaches *beyond* the locality London – the delinking of locale and milieu.

Consequently, as far as its spatial order is concerned, the London dweller's milieu, which integrates different places and people across the global city into a configuration of action and experience that is both practicable and meaningful for the individual, would appear as a *miniature of the extended milieu* that takes in relevant places and significant others across the globe. This parallel can be further illustrated by several features.

Distant places and significant others within the urban dweller's milieu are linked through metropolitan means of transport and communication. Just as e-mail and aeroplanes help to maintain extended milieux across the global distance, underground and telephone are crucial for maintaining a milieu within the London area. Sarah mentioned in this regard that 'home addresses aren't actually terribly relevant, it's phone numbers that are important'. She also explained that she would meet friends in what she calls 'third territory', instead of socializing in each other's houses, which might be too far away to meet conveniently after work. This importance of access to London's transport facilities is also mentioned by Ira, the 'metropolitan':

> Part of that thing about friendships in London is, you know, the proximity thing in a way, that you may kind of find people that make the effort. I tend to spend a lot of time on my own because of that proximity thing. I haven't particularly made any friends in Brixton [her place of residence], a lot of my friends are scattered all over London, and sometimes I'm too lazy, or I haven't got the money for transport to see them often.

There is a second point implicit in Ira's statement. Even though it might mean having to spend some evenings on her own, she prefers to socialize with people of her choice, not people who happen to live in the same locality. With the inexhaustible sociocultural variety of London at their disposal, and with access to a relatively efficient metropolitan transport network, there is potential for associating with the like-minded, in turn making involvement in local relationships less likely, or less necessary. We can also recall here that for Sarah, for example, her place of residence is simply a good 'base' for spending weekends in Barcelona and France, or alternatively socializing with friends in a 'third territory', rather than the local pub. Thus she finds that, 'I am just never there in daylight when one meets the neighbours'. Looking at it from the perspective of those who would like more local involvement, Herbert, the 'metropolitan local', regretfully observed that 'there is not the same association with a particular area and the same road anymore'.

Finally, London dwellers have to incorporate *generalized settings* into their daily lives, from grabbing something to eat in a fast-food outlet to extensively using the standardized facilities of metropolitan means of transport. As mentioned before, much of the social life takes place in a 'third territory'. Accordingly, routine meetings at an underground station – 'at the ticket barrier' providing the most commonly used meeting point – are quite normal for the beginning of both informal and more formal social activities. This is in a routine praxis that can be adapted and repeated almost anywhere in London. Moreover, many Londoners reappropriate standardized facilities as part of their milieu. Routines 'normally' associated with home, such as eating and personal hygiene, are conducted while on the move with London's public transport. It is not an uncommon sight on London's underground that someone would heartily bite into a takeaway meal, or put on make-up and file nails, or try to have a nap, completely oblivious to their surroundings. This resembles very much what Harold, the 'global businessman', told us about his ways of creating personal space on aeroplanes in order to get on with both 'desk' work and catching some sleep. Most Londoners would possibly agree with Harold's statement that 'travel is a trick, OK'.

However, the finding that milieux in large cities resemble in their spatial organization extended milieux, is a feature that is not exclusive to the global city.

In this regard London is only gradually different from, say Manchester or Dortmund. It is the content of their huge urban landscapes that makes the actual difference between the large city and the global metropolis.

Milieu and metropolitan variety

In his writing on the cultural function of the world city, Mumford (1991: 639f.) considered London to be 'the most complete compendium of the world'. At the same time, he sees the challenge for the urban dweller held in this development of metropolitan culture:

> Though the great city is the best organ of memory . . . yet created, it is also – until it becomes too cluttered and disorganized – the best agent for discrimination and comparative evaluation, not merely because it spreads out so many goods for choosing, but because it likewise creates minds of large range, capable of coping with them.

Both the city as an innovative environment, and the city as a seedbed of anomic tendencies are implied in Mumford's account, describing an ambiguity that has continued into the descriptions of the contemporary city as the embodiment of the postmodern condition. Harvey (1993: 66, 292ff., 299ff.), for example, sees the new postmodern urban spaces as liberation from the 'tyranny of the straight line', while at the same time these spaces can be described as a world of simulacra, eclecticism, ephemerality, escape and disruption.

The fate of the individual who is exposed to metropolitan variety and ephemerality, and expected to make sense of it in daily life, has been a long-standing issue in modern social sciences, most prominently taken account of in Simmel's essay concerning 'The metropolis and mental life' (1969). Although the globalized world city London of today is not the same as the nineteenth-century metropolis London, there is continuity with regard to the sheer variety and ephemerality of its social and cultural environment. Londoners can certainly relate to some of the features identified by Simmel as characteristic for the metropolis.

Most importantly, it is the sheer amount of culture, the 'overwhelming fullness of crystallized and impersonalized spirit', in Simmel's words (1969: 422), which is on offer in the metropolis, and that sets it apart from other big cities. Comparing London and Hamburg, where she has lived for a while, Sarah states in favour of London:

> Culturally there is just more here, there is more opportunity to do things. There is so much happening in it, and whatever you want to do you can do it, there will always be someone or something you can join, you can always do something you are really interested in.

Sarah also gives evidence for the lived experience of what Simmel (ibid.: 416) describes as 'a kind and an amount of personal freedom which has no analogy whatsoever under other conditions'. She told us, for example, that she feels no pressure to become involved with the local community in the place where she lives. She also mentioned that she is able to entertain two different circles of friends for different forms of leisure without these circles necessarily having to overlap. But this metropolitan freedom also finds expression in her personal appearance:

> An advantage of London is, I suppose, that it's a less conservative society. I did notice in Hamburg that people were in general quite conservative, say if you would wear fashionable clothes or something, they would just stare at you a bit. It's much easier to get away with things like that in London . . . In London you can dress as you like, actually, and you can dress quite eccentrically if you want to, or very fashionably, whatever, and people don't remark on it, no one will raise an eyelid about it. I personally feel there are certain ways of dressing which maybe I wouldn't dare if I lived outside London. So, I mean what I did find out about me in London is I like to dress in lots of different ways, and I can do that here.

Here Sarah brings up another crucial point in Simmel's analysis. The metropolis is a place of self-actualization, an environment that encourages 'unique inner and outer development' (1969: 117). And as revealed in Rolf's story about finding the courage to finally admit to his gay sexuality, these personal transformations can go far beyond the extravagances and mannerisms of fashion, right to the core of the individual's milieu. His decision to come clear with himself and others, and then subsequently to restructure his life rather drastically, was by his account triggered by the daily experience of 'other' life styles, something that just was not present in the small town in southern Germany in which he lived before coming to London. So, the illustrations so far suggest that the metropolis certainly stimulates people's milieux. It is a challenging environment that encourages the extension of people's milieux in the meaning of its reflexive restructuring.

However, the most commonly known features of metropolitan life that Simmel's analysis singles out refers to stimulation in a rather ambivalent way. The possibility of a 'boundless pursuit of pleasure', Simmel argues, leads to an 'intensification of nervous stimulation', in turn resulting in the metropolitan's 'blasé attitude' (ibid.: 410, 413). Some of Simmel's account is apparent in Rolf's narrative concerning the pursuit of his music and singing hobby:

> I mean the sheer number of concerts on offer is just amazing. So at the beginning I thought, Oh, great, I will be spending all my time going to all these concerts. Now? I don't go to anything at all anymore, because I just

don't know how to choose, which to attend. I mean, just look at 'Time Out', it's just crazy, you could go to 10 or 15 concerts per day, and then the choice is so overwhelming, you just start to ignore it all. Meanwhile I only go to things which friends or colleagues recommend to me.

One could now argue with Simmel that Rolf has developed a certain 'matter-of-fact' or 'blasé attitude' as the result of some kind of overload of stimulation of interest. Another way of looking at these developments would be to say that the blasé attitude is something that is intrinsic to the perspectival and filtering function of the milieu. It might be more evident in an metropolitan environment, but is not an outcome of it.

Schütz (1967: 208f.) terms this attitude of naturally being blasé about certain things in order to get on with the practically relevant tasks, the *natural attitude of everyday life*, by which he asserts that in the normal course of events our daily lives are guided by a 'pragmatic motive'. Now, abruptly changing the environment we move in clearly disrupts this normality. Similar to the crisis of 'symbolic consciousness' in the process of the spatial restructuring of the milieu, discussed in the previous section of this chapter, it can be assumed that there will be some sort of crisis of this 'natural attitude of everyday life' during the course of regaining a practical milieu. Being placed into a new environment is a situation that does not immediately allow for the 'normality' of established routines.

Harold's account on this issue provides illustration of the temporary character of the transition from an initial period of opening out the milieu in terms of exploring and experimenting, until finally the milieu contracts back to the routines that are necessary in order to manage both everyday life in London and commitments across distance.

> When we first came here, every weekend we explored London, OK. That was exiting, that was new, that was big. And we walked everywhere, we walked through the city, we walked through St Katherine's Docks, we went to exhibitions, we went to see everything, you know. But then you suddenly realize, wow, I own this house, I have to paint the windows, I have to do this and that. And it has to be done with some timing, because we also have a house in the US that we go back to every year in the summer, and we have to take care of this too . . . So, after a short period of adventure, unless you develop some *real* interest in continuous shopping in the markets, street markets or antique markets, I think coming into [Central] London tends to be not so exiting [my emphasis].

Living in what Schütz (1967: 229) describes as the 'epoché of the natural attitude' of our milieux, means we bracket out, both consciously and subconsciously, the variety and complexity of our surroundings. If one is to follow Schütz's argument, then experiencing the metropolitan environment through the 'alphabet' of our milieu, which is providing some sort of configuring 'filter', we are simply not aware of all the variety that metropolitan life has to offer. As with any other environment, the milieu 'organizes the [metropolitan] world for me in strata of major and minor relevance', but also 'complete ignorance' (ibid.: 310). In other words, it is perfectly possible to live in London without knowing or wanting to know, for example, about the Chelsea Flower Show, unless one happens to be someone who is really into flowers.

Once a 'natural attitude of everyday life' is re-established, the new environment and its sociocultural variety seem to find a convenient structure that is manageable in the individual's milieu. Everyday life has fallen back into a 'normal' time–space trajectory that reflects the individual's *real* interests', as Harold aptly phrased it. It might then appear as blasé when Harold outlines his interests in relation to London's cultural variety with the confidence of someone who knows what is manageable within the time–space projection of his milieu.

> My son and I came in [to Central London] to the big computer show, that's of big interest to us, whereas the Notting Hill Carnival or the New Year's Parade isn't. We would also like to go to the boat show because we like boats, and we've never been to a boat show. But again, it's gonna be here next year. There is such a tremendous amount of things to do here, probably more than in any other city, apart from New York and Chicago maybe . . . But we are in no rush at all, because we'll be here until, you know, until I retire at the end of my career. There are things to go to every year, so we have time to see them. If we see them, we see them, if we don't, we don't. I think *this is our attitude towards all that* [my emphasis].

Rather than being overwhelmed by it, Harold sees the metropolis and its variety at his disposal for nourishing a selective set of '*real* interests', which re-crystallize after an initial period of excitement. Coming back to Simmel's argument concerning the individual's positioning in the (excessively) stimulating metropolitan environment, a different way of looking at it seems plausible. What appears as being blasé in the metropolitan dweller's attitude towards the world is only the 'natural attitude' appropriate for the metropolitan environment. The filtering, and thereby configuring function of the individual's milieu is always there. It is the complexity of the metropolitan environment that makes this structuring more obvious.

If one follows Schütz's line of argument so far, then the 'intensification of nervous stimulation' is not quite as overwhelming and random as it would appear from Simmel's argument. Instead, this stimulation itself is filtered through the milieu. Scheler (1973: 144, 147) points out in this respect that 'all possible attention oscillates within the frame of the direction of interests' that defines the general disposition of the individual's milieu. And indeed, we saw Rolf not complaining about a variety of new interests in paintings, flowers and football that he had developed since arriving in London. His sense of being temporarily overwhelmed referred to the variety of options that he suddenly faced with regard to his long-standing interest in classical music. Harold, in turn, pointed out that in the initial period of excitement his attention 'oscillated' towards London's street and antique markets, but he could not really get himself to develop a '*real* interest' in them. As Scheler would point out, different environments might force us to extend the horizon of our milieu, but will hardly brake the milieu-structure.

The analytical point of interest here is that the metropolitan environment kind of 'reveals' the individual's milieu. While global mobility and life across distance make us realize the difference between 'locale' and 'milieu' as two different foci of the individual's field of action and experience, the metropolitan variety displayed in a place like London in turn brings to the fore the difference between 'environment' and 'milieu'.[3] Something which goes rather unnoticed in the rather stable routines of life in a village or a small city, surfaces in the temporary feelings of excessive stimulation and the seemingly blasé attitude: we experience our surroundings through the filtering and at the same time configuring medium of our respective milieux.

Living in the metropolitan variety of a world city reveals the individual's effort to actively generate and maintain a milieu in an environment that otherwise, in its sheer size and complexity, escapes their senses and the scope of their actions. To maintain the natural attitude of everyday life by developing ways that might be read as a blasé outlook on the world, is an indication of this ongoing project in the metropolitan dweller's life-world.

8

THE GLOBALIZED WORLD CITY
AND ITS 'COSMION'

Sir, when a man is tired of London, he is tired of life; for there is
in London all that life can afford.

(Dr Johnson, cit. in Tames 1992: 102)

Dr Johnson's remark on the eighteenth-century metropolis still holds true for today's London. The globalized world city is a micro version of the world in terms of the ethnic, social, religious and cultural variety that are integrated in its life-world, and the multiplicity of different life styles that are juxtaposed in its everyday life.

As the 'peopling of London' (Merriman 1993) shows, the integration of culture and social practices from different parts of the world into the metropolis is a long standing process, going back to its beginnings. However, globalization processes have not left the make-up of the metropolitan life-world unchanged. The collapse of spatial barriers has led to greater mobility of global flows of images, people, and social practices. With the global city being at the core of a new global cultural landscape, its life-world experiences a process of microglobalization. This process was defined earlier as the compression of global flows into the metropolitan environment without those flows necessarily being fused in a 'melting pot'.

This scenario raises questions towards the symbolic and social order of the globalized world city. Does microglobalization imply the de-regulation of the metropolitan life-world, meaning its opening up towards the global cultural economy, just as the financial sector of the former metropolis experienced its 'big bang' while being transforming into a truly global City? Does the collapse of spatial barriers mean that different social worlds simply collapse into each other?

The globalized cosmion

We get an analytical perspective on these issues within the metropolitan life-world when we take into account that the world which we encounter in metropolitan everyday life is not just an agglomeration of unrelated 'things'. Like any other environment, it is first of all a fragment of the world of everyday life, in which things are linked in structures of meaning reflecting our practical interests. Schütz (1967: 133) points out with regard to this symbolic dimension of the life-world:

> It is a world of culture, because, from the outset, the life-world is a universe of significations to us, i.e., a framework of meaning (*Sinnzusammenhang*) which we have to interpret, and of interrelations of meaning which we institute only through our action in the life-world.

This means that because of their immanent 'appresentational function', objects and events are, from the outset, 'cultural objects' and meaningful 'actions' respectively (1967: 328, 348). The world of everyday life as a world that is 'permeated by appresentational references', is subsequently described by Schütz[1] as a *cosmion* (ibid.: 336, 355), by which he means a self-articulation or self-interpretation of a social group that shared a sociocultural environment:

> A cosmion . . . is a whole a little world, illuminated with meaning from within by the human beings who continuously created and bear it as a mode and condition of their self-realization. It is illuminated through an elaborate symbolism, . . . and this symbolism illuminates it with meaning insofar as the symbols make the internal structure of such a cosmion, the relations between its members and groups of members, as well as its existence as a whole, transparent for the mystery of human existence.
>
> (1967: 336)

Schütz's definition stresses the inwards looking and integrating character of such a symbolic discourse. The underlying assumption of such a cosmion is that of an 'in-group' sharing a 'common social environment' in which members of the group can take for granted that other members of the group will interpret the world in typically the same way (ibid.: 326ff.). This in turn assumes that there is a shared common 'stock of socially approved knowledge', based on a shared system of practical relevances. As Schütz (ibid.: 338, 348ff.) insists, this shared stock of knowledge is 'of highest importance for the establishment of a universe of discourse' insofar as it provides the essential 'interpretational scheme' which guarantees that members of the group 'see' things in the same way.

Could the life-world of the globalized world city, London, be grasped as such a single cosmion of self-interpretation? To begin with, it would seem that the notion of the metropolis, as in Mumford's understanding of the world city as the most complete compendium of the world, sits quite well with the idea of the cosmion. Mumford (1991: 638ff.) clearly stresses the assembling, unifying and transforming 'mission' of the metropolis, making it an institution that is able to 'hand on to the smallest urban unit the cultural resources that make for world unity and cooperation'. Simmel, too, stresses the typicality of the metropolitan way of life, though it might exist in thousands of variations (1969: 410f.). However, this integrating function of the metropolis was expressed most clearly by Wirth (1969: 61) when he wrote about the metropolis as an 'orbit' which has 'woven diverse areas, people, and activities into a cosmos'.

From this perspective one could now argue that recent processes of time–space compression, with different sociocultural spaces collapsing upon each other, have, if anything, intensified the metropolitan dimension of cities like London. In this sense the globalized world city could simply be seen as a continuation of the modern metropolis, the most complete cosmion, and truly a 'whole little world' in itself.

However, at the same time it can be argued that the diversity of global culture has been re-structured in a landscape of global cultural flows (Appadurai 1992). In the light of what Harvey (1993: 295f.) describes as the 'central paradox of space and place', the globalized world city can then be regarded as a nodal point, or a switchboard, of these sociocultural 'scapes', rather than simply being a melting pot of global diversity. This would imply that the symbolic order of the global city's life-world is as 'exploding' as it is integrating. Evidence for this can be encountered in the normal course of London everyday life.

Imagine you are walking down London's Balham High Road towards the underground station. Your gaze is caught by graffiti that show a bearded and bespectacled man, encircled by the message 'Move heaven and earth to defend the life of chairman Gonzalo'. The 'interpretational scheme' of your stock of knowledge starts to operate: Gonzalo – Spanish – Philipe Gonzales – leader of Spanish Socialists – handsome media image of the Spanish Socialist – guerrilla features of the man in the graffito. . . . But you stop thinking about it as you get on with the routine requirements of everyday life in London on a Monday morning, i.e. find a ticket machine, buy a weekly travel card and newspaper, and rather quickly this graffito becomes irrelevant. Days later you get some context of the graffito by stumbling across one of the same style, this time demanding 'Support the people's war led by the Communist Party of Peru'.

Looking at it from the perspective of Schütz's argument, I am the unaddressed accidental onlooker, experiencing this graffito in my field of experience, yet unable to relate to the symbolic discourse in which it is embedded. In technical terms, I can identify the 'appresenting object' as a graffito, and also the 'appresented' phenomenon as the chairman of a party whose life is under threat. But I find myself unable to adequately realize the overall 'appresentational situation' in which they would make sense (1967: 297ff.). Only the accidental encounter of the second graffito gave me the 'referential scheme' towards 'the people's war in Peru'. What I miss is the 'contextual or interpretational scheme', which, we have to recall, was 'of highest importance for the establishment of a universe of discourse' (ibid.: 338).

It would appear, then, that these graffiti in their symbolic reference[2] are not directed towards me, or most other Londoners. It aims at a fraction of Londoners somehow connected with the people's war in Peru. Its symbolic reference is linked to a transnational symbolic space that links (some) Latin American People in London with like-minded people in Peru and elsewhere. Taking up Appadurai's idea of global 'scapes' it would seem that this example illustrates global 'ideoscapes' (1992: 299), defined as flows of images that 'have to do with the ideologies of states and the counter-ideologies of movements explicitly oriented to capturing state power or a piece of it', with 'freedom' functioning as the integrative image in this particular case.

London offers a plethora of phenomena that in their symbolic function are organized in similarly outwards-looking symbolic spaces, rather than integrating the life-world of all Londoners in a cosmion. The following example shall further illustrate this point, but also serve to highlight Appadurai's (op. cit.: 296f.) claim that these transterritorial symbolic flows create 'imagined worlds' that are 'deeply perspectival constructs' in relation to which the individual actor is 'the last locus'.

Imagine, you hear the familiar sound of a two-stroke engine in your residential street in South London's Tooting, which you can easily identify as belonging to a 'Wartburg', the bigger cousin of the 'Trabi', but not sharing its fame as a global media image of East Germany, often reproduced in reference to the fall of the Berlin Wall in 1989, and having achieved almost iconic status on the cover of U2's album *Achtung Baby*. In this case the 'appresentational situation' is clear, including its 'contextual scheme'. You go down stairs, and immediately you spot this Wartburg, a few houses down the road. Just a pity that you cannot share your joy about the symbolic significance that this car has in your milieu. For the other residents or passers-by, it will just be the 'apperceptual scheme' that will just about allow them to classify this 'Wartburg' as a 'car', if at all they take notice of it. Any further appresentational or contextual reference will pass them by.

The same day you meet the owner of that 'Wartburg', who turns out to be an Englishman who has worked in East Germany for five years. Whenever the two of you now meet, you joke about this Wartburg and exchange stories that are related to it. Slowly you establish a discourse that draws on the shared experience of life in the former GDR (German Democratic Republic). Walking down Balham High Road towards the underground station, the two of you talk in a mixture of English and Eastern German slang, nothing to do with life in Balham or Greater London, and presumably inaccessible for those who hasten down the tube entrance with you. We have established a 'deeply perspectival' symbolic discourse that is fed by an 'ideoscape' consisting of images and colloquialisms that refer to life in the former East Germany. This ideoscape is carried by people like the Wartburg-driving Englishman, who is participating in an 'ethnoscape' (Appadurai 1992: 297) that developed only after 1989. Stretching the illustration of Appadurai's concept perhaps a bit too far, the Wartburg itself, and the occasional Trabis on London's streets, might finally stand for the related 'technoscape' that stretches from Central Europe into the globalized world city.

Some implications arise from these short glimpses into London's cosmion. While writers like Mumford and Harvey stress a scenario that sees global spaces with all their cultural content collapsing into the cosmion of the metropolis or world city, the argument of a newly structured landscape of global cultural flows, as suggested by both Appadurai (1992) and Lash and Urry (1994), encourages an attempt to also account for the opposite tendency – namely to look at the globalized world city's cosmion as a symbolic universe that is 'constantly exploding'.[3] However, this is not to be understood as an undirected and random explosion of symbolic frames of reference, as this explosion is directed through global ideoscapes and related landscapes of sociocultural flows.

This then also sheds a different light on Simmel's notion of 'excessive stimulation' as the core element of the metropolitan condition. In this context it was the 'unexpectedness of onrushing impressions' that created the 'psychological conditions' typical for the metropolitan dweller (1969: 410). What in today's globalized world city might be experienced as 'excessive stimulation' derives from something more sophisticated than the mere sensual impressions emanating from the density and variety of cultural artefacts and practices in its everyday life-world. Much rather it is the frequent encounter of cultural artefacts and practices which in their appresentational and contextual situations do not refer to the 'here and now' in which they are physically located, but instead direct our attention towards contexts of meaning and action beyond and outside the life-world of the globalized world city.

What this amounts to is a delocalization of symbolic reference. This means first that the appresentational situations in our immediate local surroundings invite us to follow their referencing towards happenings somewhere else in the world. The Peruvian graffito at Balham underground station might act as an example. It means that perhaps, more often than not, in our encounters with appresentational situations in the globalized world city we will not be in the possession of the appropriated 'contextual or interpretational scheme' (Schütz 1967: 299). Unless we happen to participate in the particular symbolic discourse under question as far as our everyday life in London is linked to it through biographical relevance. The encounter of the Wartburg and the subsequently developing symbolic discourse on life in former Eastern Germany can stand as illustration for this.

For the Londoner that means that he or she is not participating in an in-group related coherent cosmion, shared with all Londoners. Instead he or she will have to manoeuvre a way in the landscape of symbolic discourses that coexist in their attempt to provide a symbolic order. The appropriate attitude that might be developed in these circumstances can best be described as an unconcerned attitude. In contrast to Simmel's blasé attitude, referring to the bracketing out of an overload of sensory impressions, the unconcerned attitude describes the individual's ability to follow symbolic references towards distant places and happenings only so far as they are of concern with regard to the symbolic situatedness of their respective milieu.

An unconcerned attitude towards the symbolic complexity of their London life-world is partly explained by the fact that people whose lives are conducted in extended milieux are themselves, to different degrees, part of the landscape of global cultural flows, instead of being passively exposed to them. Moving in transnational spaces, these milieux will acquire a stock of knowledge that reflects a mobile and delocalized biography. Consequently, the 'contextual and interpretational scheme' with which they approach London's complex cosmion will not be locally derived from being socialized into an in-group, but will nevertheless be regarded as 'socially approved knowledge' (Schütz 1967: 299) in certain 'sociospheres', and handed down via global 'media-' and 'ethnoscapes'.

We can illustrate this point by referring to Barbara's description of her rather uncomplicated entry into London's sociocultural environment. With regard to the more general interest of a temporary Londoner for the next two years, the contextual and interpretational scheme she applied to London's sociocultural variety refers to clusters of media symbols and tales of other American expatriates with London experience, with whom she spoke before flying here.

> They enjoyed it, going around seeing the sights. So most people said, 'Oh yes, its a wonderful place to go, you will find it very nice'. So it was really like bringing to life some of the pictures in my mind that I have from those descriptions of the wonderful Westminster Abbey, the magnificent Houses of Parliament, and the Tower of Big Ben . . . And then going to the Wimbledon Tennis Tournament, that was really fun for me, that's something we've been admiring from a distance. I like to watch Wimbledon tennis and it's really nice to be able to say 'I've been there for a day, I've seen those people play who were winning the championship'. Just to see what it looks like, and it's quite different to what you think, since on TV you see the tennis court and the stadium, but that's only one out of 14 courts . . .

Participating in the symbolic discourses evolving around the landscapes[4] of global cultural flows also allows people to maintain the structure of practical relevances of an extended milieu. In one of the previous chapters we described Barbara, for instance, as being quite alert towards all 'signpostings' referring her to the 'Filipino experience', be it the affinity for people with 'some kind of history with the Philippines', be it buying certain herbs in China Town, be it the importance of occasionally preparing and sharing a Filipino dish at home in Streatham, or be it finding out about festivities of London's Filipino community. All these things are of symbolic significance for maintaining the link between Manila and Filipino London as the crucial axis of Barbara's symbolic territory of the self.

Participating in symbolic discourses evolving around transnational cultures is another encouraging factor for maintaining an 'unconcerned attitude' towards the sociocultural variety of the world city. Someone like Harold, the 'global business-man', is in a position where he can pretty much ignore the local symbolic discourses. For him the 'London Season', for example, with the New Year's celebrations, Wimbledon tennis, Notting Hill Carnival, and so on, does not exist. By his own account, he would never attend these mass events. In the way he judges the symbolic significance or relevance of events like that he is related to a context of meaning that is embedded in transnational space of American expatriates and 'England returnees'. This is expressed most clearly when he talks about London's theatre land and the musical-crazy American tourists. A visit to any of those musicals, although London's theatre land is on his doorstep, to him would make sense only in reference back to parties of 'London returnees', where, in his opinion, lies the centre of the discourse on London's musical scene.

> They [American tourists] don't live in London, they don't have that opportunity. A lot of times they don't live in a cosmopolitan city that offers 50 theatres, OK. So they come here, they come here once, they want to see all the West End shows they can see. Because that's probably the conversation for the next 10 years at home. For us, I mean if we went to see *Phantom of the Opera*, how many people would we tell about it? You certainly wouldn't have a party here to discuss how many shows you have seen, like you would in America. If you return to America, then, I mean everybody would want to know if you saw *Phantom of the Opera*, *Miss Saigon*, *Cats*, or whatever.

In technical terms, what this narrative sequence suggests is that Harold is not hesitant to approach London's cultural variety through an 'interpretational scheme' in which certain objects and events get their contextual meaning only when related back to a social group and its symbolic discourse, which are situated outside London. As long as he is to stay in London, going to a musical literally does not make sense for Harold, as the symbolic capital he acquires by watching '*Cats*, or whatever', can only be realized in a social network in the States, to which he has no immediate access at the moment ('how many people would we tell?'). However, Harold is not alone in this way of looking at London's theatre land. This unconcerned attitude has the status of 'socially approved knowledge', insofar as it is shared with a certain group of expats with whom he socializes in London. Small talk in this circle is more likely to evolve around the issues of boat shows and computer exhibitions than musicals ('You certainly wouldn't have a party here to discuss how many shows you have seen').

138

Coming now back to Schütz's idea of the cosmion as a relatively coherent symbolic universe based on the socially approved stock of knowledge, and shared by a social group, it appears that the globalized world city's cosmion is rather fragmented and loose. There is no one unifying cosmion that would provide a 'relative natural view of the world – that is London', to play on Schütz's terminology (1967: 348). There is no one typical way of 'seeing' things for 'the' Londoner. But there is more than one way in which the 'actual configuration of the environment' is 'mapped out' and 'presymbolized' by different social groups (Schütz 1967: 348).

To be clear, the above argument refers to more complex 'appresentational situations' or symbolic references of a higher order. It does not question Schütz's assumption about the 'idealization of the congruency of the system of relevances', by which he asserts that for the normal course of events in everyday life 'we' will interpret relevant phenomena 'in an empirically identical manner, i.e. sufficient for all practical purposes' (1967: 316). We will certainly identify the same object as a London cab, and another one as underground station, and we might even find London's old ladies politely introducing us to the 'proper' way of queuing for London buses. Many other examples of that kind could illustrate this very basic stock of knowledge that most Londoners would share and apply as an interpretational scheme in basic everyday life situations. But for all deeper and complex appresentational situations, we have to allow for the possibility that there is much less overlap of our interpretational schemes, including the possibility of a 'complete disparity', making the establishment of a universe of discourse 'entirely impossible' (1967: 323).

In the light of the argument developed so far it seems plausible to suggest that the life-world of the globalized world city is interspersed with a multiplicity of cosmions[5] that link up fragments of London's everyday life-world with the landscape of global cultural flows. They can be seen as juxtaposed fragments of globalized symbolic discourses that provide alternative styles of symbolic order within the life-world of the global city. They are not necessarily bound to interact, but instead more likely to overlap. In this they reflect the 'complex, overlapping, disjunctive order' of the 'new global cultural economy' (Appadurai 1992: 296).

In this context of a fragmented cosmion then, symbols, instead of having an integrative function for shared local practices, provide *Sinnklammern* (linkages of meaning/sense; see Srubar 1988: 241) for keeping together fragments of globalized discourses 'here' and 'there'; for example, the *Sinnklammer* that link parts of Brixton's everyday life with Kingston Town, be it via music, clothing, stickers, or food. This symbolic link is certainly more prominent in the street culture around Brixton tube station, than, say, links between Brixton and its immediate local neighbour, Streatham.

One indication of the global character of these different symbolic worlds that make up the globalized world city's cosmion is their immediate responding 'here' to happenings somewhere else 'there' in the world. A final illustration shall serve to clarify this point. On Saturday, 23 September 1995, the news reported on the miracle of the milk-drinking statues of the Hindu gods Ganesh and Shiva. The event allegedly started the day before in Haridwar in India and spread from there almost in no time to the Hindu communities all over the globe. You decide to go to your local newsagent in Balham to buy a newspaper in order to read more about this, in your opinion, rather strange event. Rather by accident, the newsagent strikes up a conversation about it, and you express your disbelief. And all of a sudden you are invited to come upstairs to his family's shrine to go and see for yourself. You are left baffled to find yourself in the middle of something that seemed so far away, if not on another continent then at least as far as Southall in another corner of London. It did not occur to you that the local newsagent, with whom you maintained small talk about football and books, would turn out to also be related to another 'world' to which you have no real access. Afterwards the newspapers would clarify that this simultaneous appearance of this phenomenon across the global landscape of Hindu Diaspora was not really a miracle, but an indication of the complementary power of 'ideoscapes' and 'mediascapes' (*Independent*, 23 September 1995: 3).

In the light of these illustrations of coexisting symbolic worlds in the globalized world city, which more or less immediately reflect happenings somewhere else, it seems appropriate to suggest that these overlapping and criss-crossing symbolic discourses in the metropolitan life-world form a *prism* rather than cosmion. The global city's life-world is a prism that reflects and refracts the world's landscape of cultural flows.

Moving in sociospheres – the 'unknown others'

The city is, as Sennett (1993: 48) remarks, 'a milieu in which strangers are likely to meet'. This holds true even more for the globalized world city at the core of a new geography of global culture. Time–space compression and complementary processes of individual and collective mobilization have been eroding the relatively stable ethnogeographical order of previous centuries, in which the 'other' had an allocated place in a fixed spatial order. The 'other' now appears in immediate vicinity, no longer as an exception in relatively isolated social worlds, but as the rule in an increasingly heterogenous world of everyday life. This observation can be made in many social environments, but is certainly most obvious in the globalized world city, the everyday life of which is at the centre of the central paradox of space and place that drives the restructuring of the new global cultural economy (see Lash and Urry 1994).

However, the 'stranger' or 'the others' coming into the globalized world city do not immediately determine how they relate to each other. It can be assumed that the way these 'others' relate to each other will reflect the mobile and disjunctive character of London's cosmion. Harvey (1993: 301) suggests in this regard that a strong sense of 'the Other' is replaced by a weak notion of 'the others', 'reflecting the contingent and accidental "otherness" in daily life'. In the following I want to underline this suggestion by discussing two classic ideal types of social constellations between 'others'. Schütz's idea of the 'stranger' and Sennett's notion of the 'cosmopolis' will be discussed against narrative material.

Schütz's concept of the 'stranger' is closely related to his idea of the cosmion, discussed in the previous section as the communicative environment of an in-group whose members share the same interpretational scheme for going about the business of everyday life. The ideal typical figure of the stranger is now someone who appears in a constellation between 'home group' and 'foreign group'. Unlike Simmel's (1969: 402) stranger, the 'wanderer' who 'comes today and goes tomorrow', Schütz's (1971: 91, 97) figure of the stranger reflects the 'newcomer' and 'immigrant' who attempts to become permanently part of the group she or he approaches. To achieve acceptance, the 'newcomer' has to go through a 'process of social adjustment' at the end of which the 'newcomer' will cease to be a 'stranger'. What starts as a crisis of social competence, insofar as the old accepted assumptions do not work any longer, finishes with the 'newcomer' having taken over the socially approved stock of knowledge of the in-group (1971: 96ff.).

Sennett, on the other hand, develops an alternative typology of the stranger against the background of the eighteenth-century 'cosmopolis'. He differentiates between two types of strangers, the 'outsider' and the 'unknown' (1993: 48). The stranger as outsider belongs to a neighbouring ethnic group and can be clearly allocated in the city's ethnic and cultural landscape. The outsider, in Sennett's typology, is, on the one hand, clearly identifiable as 'different' through skin colour, language or cuisine, for example, but is safely embedded in the social world of his or her group, and thus does not pose any real threat towards the identity of other ethnic groups. Sennett refers to New York outside Manhattan as an example for this type of constellation (ibid.: 48f.).

The second type introduced by Sennett is the 'unknown' stranger, who is not identifiable by his or her social qualities, and therefore not easily allocated within the city's sociocultural landscape. Accordingly, it is this unknown stranger who escapes 'schematization', and who potentially poses a threat to the urban order. As Sennett argues, this type of stranger is prevalent when a society is in the process of reorganizing its social universe, thus temporarily lacking basic rules of social identification and classification (ibid.: 48f., 59). Sennett goes on to describe eighteenth-century London as example of a city full of 'unknown' strangers. In the context of industrialization processes, masses of people were drawn from London's hinterland into the metropolis, which Sennett (ibid.: 51) portrays as 'enormous city, for its time, growing at least by half through the in-migration of young and unattached people'. Cut loose from their indigenous social contexts, and thus without the background of family and occupational status, they were massed together in the proximity of urban space. While deprived of basic rules of how to define and approach each other, there was need for new patterns of social interaction 'which were suited for the exchange between strangers'. Accordingly, the 'new urban society' compensated for the 'lack of sure rules' by developing forms of 'sociability' that were 'nonparticular' in character, namely, 'applying indiscriminately to any person'. Eventually this meant the necessity to generate a new 'universe of social relations' (ibid.: 17f., 62f., 80).

There is continuity and discontinuity with regard to the constellations outlined by Schütz and Sennett on the one hand, and situations that can be found in the globalized world city. To start with Sennett's idea, the globalized world city has certainly features of a cosmopolis, where unknown strangers from different backgrounds are thrown together, and have to get on with each other wherever everyday life encounters dictate. Thus metropolitan forms of 'sociability' appear to be of unchanged importance.

Yet, at the same time, globalization processes have changed the social and symbolic make-up of the world city, and subsequently altered the conditions in which strangers will potentially relate to each other. As Sennett (1993: 51f.) points out, those unknown strangers that flocked into eighteenth-century London came 'from points fifty miles or more from London, and fifty miles was at that time at least two days travel'. In other words, a distance sufficient for those migrants to be 'cut loose' from their previous social environment. In comparison, time–space compression increasingly allows for 'extended milieux' who are linked in different ways with the landscape of global cultural flows. As outlined in detail in the previous section, this implies, amongst other things, that people coming into the world city are not necessarily 'cut loose' from the social worlds and the symbolic discourses in which their milieux were previously embedded. This in turn suggests that there is more leeway for (non)engagement with other social worlds and their symbolic universes, irrespective of whether these are local or extended worlds. To simplify, we could say that globalization processes bring more strangers from more distant parts of the globe into the local vicinity of the world city, while at the same time making the social engagement between strangers less necessary.

To grasp these seemingly contradicting tendencies, Albrow's notion of 'socio-spheres' (1997: 51) seems rather appropriate. By this he means 'fields of concern or relevance' that materialize in 'networks of social relations of very different intensity, spanning widely different territorial extens'. In their local intersections these 'sociospheres' form a 'socioscape' (ibid.: 52), suggesting coexistence and overlap between these sociospheres, but not necessarily their intermingling and fusion. It can now be argued that the generation of global sociospheres and their formation of different socioscapes within the globalized world city will affect the position and perception of the stranger. The constellation of a socioscape, then, can be discussed as an alternative to the configurations of Sennett's 'cosmopolis', and Schütz's 'homegroup/newcomer'.

Extended milieux are socially maintained by participating in one or several such sociospheres. Barbara and her extended milieu, stretching between Nashville, Manila and London, can serve to illustrate this point. Having worked in the Philippines for a substantial period of time, and wanting to return there after her spell of working in London, she was keen to maintain the Filipino experience or Filipino connection. And indeed, she found relatively uncomplicated access to Filipino London, be it through socializing at work with people who 'all have a history with the Philippines', or through the Filipino friends she made at St Anne and St Agnes Lutheran Church, or via participation in events of the Filipino community. In this sense, as I said earlier, she has not moved from Manila into a completely unfamiliar environment, but from the Philippines into Filipino London. Socializing within Filipino London, Barbara is no stranger in this fragment of London's globalized cosmion. She can rely on past experience, and taken-for-granted assumptions acquired elsewhere hold true in this context. In this context she also does not encounter strangers of any type, but can relate to people that are typically familiar to her. In brief, she feels '*at home* with the Filipinos'.

Barbara's example also illustrates that a milieu can find embeddedness in more than one sociosphere. Despite her strong emotional links with the Filipino experience, Barbara is also part of the global network of American expatriates. And this sociosphere has paved another way into London for her, so to speak.

> Friends of mine in the US have given me addresses of relatives and friends, who live in London and elsewhere in England at various places, where I might write or ring and introduce myself and see if they are interested to make some contacts.

And indeed, Barbara ended up staying in the house of another American expatriate family, and she mentioned how she enjoys the socializing with other American expats at St Anne and St Agnes Lutheran Church. If anything, Barbara's case provides an example of someone moving in a constellation of globalized sociospheres. Navigating between those sociospheres, she is not necessarily bound to encounter the configurations suggested by Schütz and Sennett respectively. As long as she moves in the sociospheres that generate Filipino London and 'American Expats' London', she is neither the newcomer (Schütz), who has to adapt comprehensively to the ways of an established in-group of Londoners, nor is she the unknown stranger (Sennett), cut loose from previous social relations, status, and experience. However, the fact that Barbara does not encounter them does not mean that these constellations do not exist in the social environment of the globalized world city. All that is argued for here is that the 'in-group/stranger' and 'cosmopolis' are no longer principles of social order that comprehensively structure London's cosmion.

Herbert, the 'metropolitan local', certainly lives in a milieu that is ordered by an 'us and them' perspective that resembles Schütz's in-group/stranger constellation. He considers all those who do not accept the established 'rules' of the 'indigenous people' as 'strangers' or 'foreigners'. As a member of the indigenous in-group he feels entitled to set standards to define those 'who haven't got the *same attitude* towards living together'.

> I think, that if the numbers are kept within reason, and the people live according to your rules, speak your language, and work honestly, you know, work in a job, then, I feel you get on with them. Yet when they don't live within your rules, when they speak their own language and won't speak yours, and you can see that they are not interested in working for a living and are living on social security, then you get social unrest.

The stranger who does not comply with those rules is perceived as a threat to an established order. There is also a clear anticipation here of the fact that increasing numbers of strangers would threaten the rule-defining power of the 'indigenous' in-group. While his above statement about the stranger is clearly fed by the 'immigrant', thus inducing a somewhat racist undertone, Herbert is quick to generalize his definition of a stranger towards any foreigner attempting to settle in London or England.

> Now, I'm not the only one of my, you know, indigenous white population, who find that highly offensive that we can't venture an opinion or anything, which concerns the immigrants or the ethnic minorities without being branded racist . . . I would feel the same if they were white people, say Germans or Americans. I don't want anybody to come in large numbers, it's all a question of numbers.

So while Herbert's account gives evidence of the lived reality of the 'in-group/immigrant' configuration, we should not be surprised to find the cosmopolis type of encounter between strangers expressed in Nicos' narrative, as for the entrepreneur in London's service industry the encounter with the unknown stranger is part of daily business. His customers come from a wide spectrum of sociocultural backgrounds. And so 'if you want to please your customers, from tourists to pimps and barristers', Costa says he has had to develop social skills that allow him to relate to their 'different attitudes, different styles, different ways of talking'. He describes the game of sociability, almost like Sennett, as a game without settled rules, changing from situation to situation, and determined by the individuals engaging in it:

144

I can't really put it in words, I mean you have to be brought up in London to know all this. Even to me it's still difficult. And, of course, you can easily make mistakes, there is no rule to the game, you know. It's always a different game altogether. You know, you can't have the same conversation with somebody whose borders start and finish in Tooting, and someone who is broad-minded. You have to change your style with each individual, I suppose. I suppose that's a good thing about London, you learn to become, you learn to be a real good actor.

As much as Costa likes the game of sociability in daily business, he dislikes the uncertainty it brings to his social life. He thinks the downside of this easygoing way of relating to the stranger is mistrust, a form of protection from getting hurt and being disappointed. Costa's disillusion is expressed in particular with regard to making true friends. The argument he brings forward to underline his view once again resembles Sennett's account of the 'cosmopolis being an agglomeration of strangers cut loose from their sociocultural and professional background'.

People have learned to be mistrusting in this big city. And, I don't blame them for it. Let's face it, I meet you today, I don't really know your background in Germany, how do I know who you really are in Germany? So it's hard for people to trust. And again, this brings me back to the fact that I've said I don't really like living in such a mixed society for that reason. Because you can very easily get unstuck, because you meet people, you don't really know who they are, why they have run away from their country, while if you are back home where, you know, everybody is German, or everybody is Greek, or, you know, at least the majority are, then you can feel a little bit more at ease with people . . . It's a sad situation, but that's big city life.

However, these illustrations towards the in-group/stranger and the cosmopolis constellation as a lived experience do not counter the assumption made earlier, namely that in the light of globalized socioscapes, these are no longer the organizing principles of encounters between 'others' in the globalized world city. Moreover, it seems plausible to consider these constellations themselves as being expressions of particular sociospheres. This assumption becomes more accessible when we recall that sociospheres, as fields of concern and relevance, can have very different extensions in time–space. In their realization as patterns of social practices they generate different, that is perspectival, social realities (Albrow 1997: 53; Appadurai 1992: 296). Accordingly, each of these realities will shed a different light on the 'other'.

Looking at Herbert's account from this perspective again, it reflects the 'concern' of a milieu that is embedded in a sociosphere that is local both in scope and attitude. The whole story of Herbert moving house to local community-style social settings, outlined in detail in chapter 4, reveals something of a negotiation process between sociospheres with different social practices and different perspectives on the 'good life'.

> Where we lived in Stamford Hill, there were all white families down the street. And they were all like a street village, everybody knew everybody else . . . But then coloured people started moving in to our street. Now, there was a certain friction, because coloured people, for one thing, have a different attitude towards life, towards noise and playing music loudly late into the night. And that causes, that does cause conflict. And so therefore people started to move out, not necessarily because they object the fact of black colour, but, but it doesn't have the same discipline, say as you have.

What causes 'concern' here is not so much the spatial presence of the 'others' in immediate proximity, but their 'attitude' (towards 'life', 'noise', 'discipline') that for Herbert clearly defines the 'others' as 'strangers', in other words, belonging to a 'sociosphere' with its distinct social activities ('playing music loudly') and respective time pattern ('late into the night'). The conflict of relevances in this case obviously does not allow for the formation of a socioscape, or the tolerable local coexistence of different sociospheres, so that one sociosphere has to give way.

This encounter does not fit into a simple in-group/stranger constellation. What Herbert refers to as 'street village', 'indigenous people' or 'us', reflects the perspectival construct of a certain sociosphere, while in reality, the group described in those terms is not in the power position to set the rules of behaviour or standards of attitude for the locality in question. Both sociospheres, the one that practices localized values and behaviour, and the sociosphere of the 'deviant', that is noisy and undisciplined strangers, are not forced to engage in a process of adjustment of standards where one group would impose their standards on to the other in the attempt to generate a rather homogenous local culture. Without intending to trivialize the 'friction' and the 'conflict' involved in the process of one sociosphere finally giving way to the other, it also indicates the 'leeway' of the global city's socioscape in terms of negotiating the spatial distribution of social practices. As a de-localized field of concern and pattern of social practices, a sociosphere can evolve wherever there are like-minded people to share concern with, and to join in social activities that reflect the disposition of the participating milieux. In Herbert's particular case this meant eventually leaving the global city and settling in a village just outside London, 'By coming to this village I regained a bit of that what it was like in London in those good old days . . .'.

146

While he might have felt on the receiving end when leaving Stamford Hill and moving to a village just outside London, Herbert is at the same time clearly aware of more subtle means of securing free play for his milieu and the ongoing transformation of London's overall socioscape. Concerning the village in which he now lives, and where, up to now, the only 'coloured person' is the Asian shopkeeper, he thinks money and property will hold the flow of mobile strangers, at least for the time being. As he exclaims, 'they couldn't afford to come here, they couldn't buy a house here'.

Looking back at the argument, it seems plausible to assume a transformation of the in-group/stranger constellation itself. The existence of a socioscape consisting of relatively de-localized sociospheres, makes the coexistence of social worlds more likely than their conflictual engagement about compulsory local standards of behaviour and attitudes. While for some this means that the flexibility of their sociospheres allows them the freedom of not having to engage in such a constellation, it is for the same reason fairly difficult for others to exercise the necessary social and moral pressure for an in-group/outsider constellation to be practically effective.

Similar transformations can now be outlined with regard to the cosmopolis constellation as illustrated by Nicos's case. No doubt the globalized world city offers lots of situations in which an encounter with the unknown stranger is likely. At the same time, this does not mean that everybody living in London will necessarily have these encounters. Someone like Nicos, working in a takeaway in one of the busiest High Streets in London, is bound to encounter the unknown stranger. Someone like Harold, who spends the week travelling Europe, while returning for the weekend to one of London's quieter well-off suburbs, is, as we shall see, in the position to avoid meeting the unknown other.

Nicos, though, is aware of the fact that the position he finds himself in with regard to having to relate to unknown strangers day in, day out, is due to the profession he has grown into, but is by no means only possible one. Looking across London's socio- and ethnoscape, he knows that there is another, less cosmopolitan, way of living for a Greek-Cypriot like him, just a few miles across the Thames in North London.

> No, unfortunately, you see, I've never really lived over there [Hackney, Harringay]. I've always lived in other areas. It wasn't out of choice or whatever, you know, its just where my father took us in relation to his work, and after that of course, where I had to go in relation to my work, and it just hardly ever took me to the Greek communities.

Costa sometimes looks in envy towards this lived alternative reality, which at some point in his life was a real option, and perhaps still is. However, engaged to an English woman now, it seems less likely, and he feels sometimes quite drawn towards life in a sociosphere of extended family networks of like-minded Greeks.

> Perhaps, if I did, I would have married a Greek girl by now, which would have been nice, because, of course I would prefer that, yes. Because, I believe that, although I have a good relationship to my fiancée now, but, a person from your own country can understand you a little bit better, rather than somebody who is not from your country, plus it's to do with future family and so on, it just helps a great deal language wise, makes sure that the children learn Greek, and so on.

So while Nicos more or less has to put up with life in a cosmopolis-type sociosphere, Harold, in comparison, is in the more privileged position to negotiate his encounter with unknown strangers, which for him largely means avoiding them. For Harold the agglomeration of unknown strangers merges into an amorphous mass, a 'crowd'. Accordingly, he avoids settings in which 'crowds' develop, such as Oxford Street. On the other hand, he sees his daughter liking this kind of metropolitan atmosphere. In his account, the ability to move in certain sociospheres and to avoid others, comes down to spending power.

> For me to go shopping in Oxford Street? Never, never, never. I should avoid a crowd like that if I have the convenience to shop somewhere else. But also I have a little more money, and I don't have to go for the bargains. On the other hand, when my daughter comes back from Texas and comes here, for her it's no problem going to Oxford Street, she likes that. But also she has less money, and she is really looking for the bargains.

Relatedly, 'crowds' of 'unknown' strangers are not only perceived as unpleasant by Harold, but also seem to spell trouble. For that reason he avoids the cosmopolitan experience of the Notting Hill Carnival.

> Every year I watch it on TV, every year I see all the trouble, OK, and I think 'Do I really wanna go down there?' I look at the mess, the crowd of people, and I say to myself: 'No!', you know, 'well, that's it, it's a nice day, let's go on a picnic instead'.

Harold's account certainly displays a position within London's socioscape where he is able to consider having a cosmopolis-type experience, only to decide against it in the end. Socializing with his business partners all over Europe during the week, and with fellow (American) expats while in London, Harold is neither bound to meet the unknown stranger, nor to encounter situations where he would be one of many unknown strangers. Accordingly, he is not forced to acquire the social skills to deal with such situations, for as far as he is concerned, he will never use them. For Harold, the Londoners outside his sociosphere are perceived as the unpleasant crowd (e.g. Oxford Street) or the disorderly crowd (for example, Notting Hill Carnival). He is aware of their coexistence in spatial proximity, and the option to engage with them ('Do you wanna go down there?'). But in the end he makes sure that 'his' and 'their' London do not interfere ('let's go on a picnic instead'). There is no interest and no need to expose himself to London's cosmopolis.

It could be argued now that Harold's position is a fairly privileged one in that both spending power and access to global networks allow him to be rather aloof with the people around him and also the sociocultural landscape of his adopted London. So in order to further reveal the actual transformation of the cosmopolis aspect of the globalized world city, Harold's account can be complemented by looking at the case of Ira. We called her the 'metropolitan' because the overall disposition of her milieu is open and outgoing towards London's sociocultural variety. However, it is not only this that puts her in contrast to Harold, but the fact that she has hardly ever left London. Also, as a part-time sales assistant in a book store, and a part-time student, she has certainly not the same spending power as him to buy her way out of encounters with the 'other'. Instead, we can recall, Ira engaged with the 'other' or 'stranger' out of choice. She left the council estate in Battersea in order to 'find out about myself' by 'opening my sort of avenues to other people'. And, as she says without exaggerating, 'I love meeting people'.

So Ira, it would seem, is the 'cosmopolitan' or 'cosmopolite' straight out of Sennett's book, where he defines (1993: 17) such a mind set as referring to someone 'who moves comfortably in diversity'. That certainly applies to Ira. But Sennett goes on to say that 'the cosmopolitan, whatever his pleasure in worldly diversity, *of necessity* must make his way in it' (my emphasis). As much as Ira enjoys meeting other people, she likes to meet them on her terms. There is a certain confidence about the way she 'plays' London's cosmopolitan dimension according to her own interests. Depending on the, in her own terms, different 'vibes' she is in, she knows where to go in London in order to find a setting where she is likely to have the encounter she envisages.

> I mean, you know, I'm in that single vibe at the moment, so I have been down to the West End a couple of times recently, because it's a focal point for *giving myself the opportunity* to meet other [single] people [my emphasis].

This statement certainly does not express necessity, but competence and control with regard to London's cosmopolitan environment. This competence derives from knowing London's socioscape to an extent that generates an awareness of where different sociospheres would intersect. This awareness gives her the option to either avoid these places or to frequent them if she feels like it. With regard to London's theatre land Ira illustrates:

> You will meet lots of different people here, people are passing through it in transit, you know, you can sit in certain pubs, and meet the theatre-goer, if you want to, or the Americans, if you want to, or what have you.

Though Ira is certainly a cosmopolitan by heart, she seems to be a different type of 'cosmopolite' from the one living in Sennett's cosmopolis. Knowing the socio-sphere(s) of the 'other', allows her to determine which unknown other ('theatre-goer', 'Americans', 'single') she wants to get involved with, and to what extent (to meet, to watch).

There is a new quality in the cosmopolitanism practised by Ira. Different from Sennett's 'cosmopolite', Ira shares with the metropolitan *flâneur* the voluntary character of her encounter with the 'other'. And yet, unlike the *flâneur* who enjoys the unpredictability of the cities rhythms and the accidental bumping into the strange and the stranger, Ira knows where and when to meet the stranger or experience the strange on her terms and in settings she is able to choose and feel comfortable in. Ira could thus be best described as a *connoisseur* of metropolitan life. She is able to relate her milieu to different sociospheres containing the unknown other, and pick encounters more or less to her liking ('vibes').

It becomes also clear from Ira's account that, with regard to Harold's seemingly privileged position, encountering the unknown stranger is not necessarily the faith of the less privileged. In fact, and as described in detail in chapter 4, Ira had to uproot her milieu from the 'totally entrenched estate life' in Battersea, where she might never have encountered the 'theatre-goer', or the 'Americans'. Again, this seems to underline her description as a connoisseur of cosmopolitan life, for the connoisseur has to work on developing a taste for things and a feel for selective pleasure.

These considerations, then, seem to support Harvey's call for a weaker sense of 'the others'. In its cosmion of intersecting sociospheres, London, the globalized world city, accommodates '*unknown*' *others* rather than 'outsider' – or 'unknown' strangers. For the notion of the 'stranger' implies a threat or challenge to an established sociocultural order. With London's cosmion resembling an intersection of more or less de-localized sociospheres, there is no unifying and all-embracing social and symbolic order. The existence of globalized sociospheres also implies that the newcomer is not necessarily a stranger, but (at least typically, if not, as in the case of extended family networks, personally) known to those whose milieux participate in the same set of social activities and practices, and who share the same type of concerns and interests. Moreover, unknown others, respectively conducting their daily lives in different, (locally) coexisting sociospheres, are of no immediate concern for each other, as long as their respective 'fields of concern and interest' do not get into conflict. For most parts of everyday life, the unknown others will be aware of their respective existence, but not necessarily become involved with each other. This in turn encourages a more relaxed attitude towards accidental otherness in everyday life. The engagement with the unknown other then becomes a question of choice rather than necessity.

That this is not just theoretical talk but lived experience in the world city's globalized cosmion, can be further illustrated by narrative material. What is expressed in these sequences is an awareness of this rather tolerant coexistence of and with others. Ira, not surprisingly perhaps, likes 'her' London , because it allows the 'other', whether 'outsider' or 'unknown', to be maintained rather than absorbed or dissolved.

> That's why I think London is quite unique, because it, London really had to accommodate its foreigners, you see, people have come here to live, also people who are noticeably not from this country, you know, and it has allowed those people to have their kind of bit of London, if you see what I mean. They are allowed to have their own films in the cinemas, for instance, you know, for that culture to remain. And some people would say that's a bad thing, but I don't think it is, you know, a little bit of the Caribbean, a little bit of Turkey, or what have you, I think that's wonderful for a city like this . . . One thing that really brought it home to me the other day is like, you know, if you go to a hospital now, some hospital, you will find all different languages, like Urdu and Punjabi, and Greek, and what have you, you know, you will find these things here, and that's brilliant, *because London has accommodated them all* [my emphasis].

Ira does not feel threatened in her own milieu by the coexistence of other socio-spheres. Her own experience convinced her that 'London can take care of so many people's needs'. And this loose hanging together of different ethnicities, life styles and cultures is not something that is restricted to London's entertainment district, but something which she sees to a higher or lesser degree elsewhere in the metropolis – in the service sector (hospital) or her residential area, Brixton.

> I like, you know, all sorts of people hanging around together, and I think in a way there is something for everybody in Brixton, for all kind of different life styles, yes I like that.

What comes also through in Ira's account, is that she has no desire to absorb or incorporate these fragments of different cultures and life styles into her milieu. 'I like *a touch* of everything in my life' as she says. Looking at it the other way round, Ira is not afraid of losing her identity in the encounter of London's sociocultural variety. Just as she does not want to absorb other cultures, she is confident of not being absorbed into a particular style or way of life. In her worlds, 'I wouldn't go into a particular style, I have my own style'.

A similar view on things is expressed by Sarah, the 'mobile cosmopolitan'. Like Ira, she believes in London's capacity to accommodate its newcomers, and trusts their ability to settle in the landscape of London's social and ethnic mix, 'Whoever comes to live here, I think people tend to fit in and find something that suits them fairly quickly'. Working during the day in the City, the coexistence with 'the other(s)' is something that becomes relevant for her entertainment and spare time activities. The other is something to choose from when she feels like it, like when on an evening out with friends 'you can choose and eat anything from any part of the world'. And Ira's Brixton is for her the place 'to go and buy exotic things from the market'. The Brixton of racial friction and latent violence she will possibly never encounter, apart from on the television, as she keeps to the market and avoids it for late night weekend entertainment. Sarah, in relation to her job in the city, her singing hobby, and her extended network of well-off sailing friends, is unlikely to have accidental or unwanted encounters with unknown others; she feels quite competent to negotiate her daily life in the socioscapes of London. Accordingly, she states in a confident manner: 'I mean I know how to get myself about, I know how one needs to operate to be successful and happy in London'.

So, where does this leave the stranger? Dissolved in tolerant cosmopolitanism based on overlapping but hardly impinging sociospheres? Sarah indicates the existence of another type of stranger that might become more prevalent in the global city with its globalized cosmion and the related technological spaces. Sarah demands from everyone living in a place like London to have basic skills in relation to what she terms as 'knowing how to get yourself about'. So for her it is not the ethnic or social background, different attitudes or habits, that define the stranger. Instead, the stranger would be someone who, in an environment of spatial and cultural co-existence, manages to technically interfere with her milieu. She gives an illustration:

> For example, today I was at the airport, dropping someone off, and obviously, an airport is going to be blocked with foreigners. But there were people trying to use the ticket machine for paying the car park fee, and they couldn't understand it, and, of course, I was in a hurry, and was standing there, you know, getting more and more ape, because these people just couldn't work out how to use this machine.

So while Sarah is relaxed and tolerant about coexisting otherness, and exotic strangeness, she is not amused about those who 'out' themselves by being what could be termed *technical strangers*. What is reflected in terms of life-world by Sarah reflects what Hannerz (1998: 127f.) has recently described as a shift in the organizing principle of contemporary 'heterogenetic cities'. In these, according to Hannerz, 'the prevailing understandings and relationships would have to do with the technical rather than moral order, which is to say that administrative regulation, business and technical convenience would be dominant'.

What appears to be of crucial importance then for the stability of the globalized world city is forms of social order that allow for *detachment* (Waldenfels 1990: 63f.), in the sense of maintaining the free space between unknown others and their respective sociospheres. In its philosophical dimension this means not attempting to actually understand or appropriate the others (Ira: 'a touch of everything', 'accommodating foreigners'), but to be aware of the time–space patterns and other technical relevances of their milieux and sociospheres (for Ira to know when and where to go to watch the Americans leaving the theatre; for Sarah to go to Brixton market on a Saturday morning, but to avoid Brixton on a Friday night). In its systemic dimension this implies the ability to competently navigate the technological spaces of the metropolitan landscape (airports, cash machines, ticket machines . . .) as they are hot spots for the intersection of different milieux and their sociospheres.

A city of 'neighbours'?

The argument developed in the previous section established that the categorical type of the 'stranger' is possibly not best equipped to describe the new social constellation of *co*-existing extended milieux and their sociospheres, as can be found in the globalized cosmion of the world city. In first approximation we described the loose hanging together of different milieux with their respective social practices and overlapping time–space patterns as a world of unknown others. The term stresses the potential for mutual awareness yet at the same time non-involvement between a plurality of different milieux and sociospheres.

Recent research in phenomenological sociology has suggested the symbolic type 'neighbour', defined as the 'nearest other' (Grathoff 1994: 29ff.), for grasping constellations as the one described above. The reasoning behind this category, at first perhaps rather curious, is that 'approved detachment' (ibid.: 46f.) between coexisting milieux is mainly achieved not through rules and regulation set by society or technical systems, but through the inter-subjectivity of the life-world. We have to recall here the definition of milieu as a relatively stable configuration of action and experience, structured by the individual's 'biographical situation' and its relevences. As such it is a field of options without strict boundaries. However, it is the life-world presence of the 'nearest other' that marks out the *Potestativität*,[6] i.e. the horizon of possibilities (ibid.: 36), of a milieu. In other words, before any other type of intervention takes effect, the 'nearest other' in his or her milieu both opens up and narrows down the options available in my milieu with regard to the spectrum of possible actions and experiences.

The way(s) in which neighbouring milieux maintain their respective optional fields cannot be fixed into a set of established social rules, but is each time a unique constellation of overlapping and conflicting relevances. Accordingly, neighbouring milieux generate and maintain shared boundaries through tacit ways of negotiating implicit and transitory rules. The crucial paradox here is the mutual care directed towards *shared* boundaries in order to maintain *separate* milieux. 'Approved detachment', the category attempting to grasp this paradox, suggests communication but not consensus and understanding; awareness and recognition but not involvement, and distance despite proximity. The nearest other helps to maintain the 'free play' of my milieu, and vice versa, but he or she is not necessarily a friend. Yet he or she is also not a stranger, for I trust in the care and maintenance work she or he invests in a shared boundary that implicitly contributes to the situatedness of my milieu.

It is now important to recall that the neighbour as the 'nearest other' describes social orders of a certain type that embraces more than the relationship between actual local neighbours. Still, the fragile and unique relationships we build up with a neighbour in the common sense meaning of the word, can serve as a first approximation towards the type of symbolic order of the life-world that is associated with the metaphor of the neighbour as the nearest other. Harold and his local neighbour can serve as an illustration of the unique ways by which approved detachment between milieux is organized and marked out.

> My neighbour is a woman, probably in her forties, unmarried. Her family own an Italian food distribution centre, or something like that. But I cut her shrubs, . . . because her shrubs and my shrubs border. So when I'm cutting my shrubs, I just go over and cut hers, at both sides, OK. It would be hard for her to cut, because it's hard. In return we always periodically see a bottle of wine pitch up on our portal, or Italian cake of some kind.

What Harold knows about his neighbour does not suggest much socializing between the two parties. Yet he acknowledges the (co-)existence of her milieu with maintenance work on their border, by cutting the shrubs. This is clearly a symbolic gesture with a lasting impression that suits his busy working life and frequent absence from London. His female neighbour returns this acknowledgement with the (periodic) indication of her appreciation of his care and maintenance work through a bottle of wine. The way Harold describes this neighbourly relationship suggests harmony, but a harmony that rests on detachment. Neither enters the other's milieu. Harold works on the shrubs, not in her house, she puts the bottle of wine on the portal, rather than handing it over in person and coming in for a cup of tea. There are no further demands made towards each other, such as having barbecues or having each other around for tea every other week. We do not know the prehistory of all that, but it certainly can be read as an implicit statement saying: that is the

154

arrangement that suits both parties, let's keep it that way. The harmonious way in which the local aspect of his milieu is ordered in tacit agreement with his neighbour certainly has an influence on the 'free play' of Harold's extended milieu. There is one thing less to worry about when he flys out to somewhere in Europe on a Monday morning, knowing that his wife is unlikely to have trouble with that neighbour while he is away. And the precious spare time he has he knows he can spend socializing with other American expats rather than having to entertain the person that happens to own the house next to him.

There is certainly a multiplicity of ways in which neighbouring milieux contribute to each other's situatedness. As will there be different ways to realize approved detachment. Harold and his neighbour have quite personalized ways of acknowledging and marking the boundaries of their milieux. This might look different in the context of a residential street in a less well-off area of terraced houses converted into bedsits, with no shrubs or other shared communal facilities to be used as playing field for the symbolic markers between milieux. If, in addition, most people work and also socialize outside the residential area, local neighbours might no more recognize each other than regular passers-by on the way to the underground station. In this case the arrangements of approved detachment between nearest others will be more general, resembling forms of 'civil inattention' (Goffman 1966: 84), that is, ways of showing appreciation of the presence of the other while at the same time indicating an interest in mutual non-interference. A situation of this type is described in Ira's account on her residential street in Brixton.

> There is something nice about having that feeling of coming home to something familiar which is not anonymous, and people know you, and they nod at you, even if they don't say 'hello' . . . I like to walk up and down my street and see the same people, even if I don't talk to them.

We have to recall that Ira mentioned before that she would rather sit at home on her own when she cannot afford a travelcard to socialize somewhere else in London with friends from work and university, instead of attempting to mix in with the Afro-Caribbean community or socialize with her local neighbours. Yet her statement shows how much she values the non-intrusive presence of certain nearest others. Mutual acknowledgement with them via nodding contributes to the situatedness of her milieu ('coming home to something familiar') in this particular street of London. Again, there seems to be the tacit agreement of leaving things as they are, of not wanting or needing to get deeper involved with each other. If, however, these invisible boundaries between her milieu and the nearest other breaks down for some reason, Ira knows how to protect her milieu by putting up barriers, or in her words, 'I put up my guard'. One way of maintaining detachment from the other who is not willing to engage in forms of civil inattention, is to play with different English accents. She gives an example referring to a group of male youths, sometimes hanging out in her street, and occasionally trying to chat to her.

> If I want to keep distance, which I do most of the time, then I would continue to talk this rather posh English, and I can see in them that they don't know how to relate to me, I mean they are really puzzled by that.

As pointed out before, symbolic orders of the type neighbour do not only apply to the arrangement between two particular local neighbours and their milieux, though it might come to the fore most clearly in these situations, but can be extended towards other forms of approved detachment with the nearest other. We saw in Ira's case the extension towards an unspecified number of nearest others, with a fairly loose 'neighbourly' form of civil inattention guaranteeing the situatedness and at the same time 'free play' for her milieu. Listening further to Ira's account on 'urbanism as a way of life', it would seem that metropolitan life is governed by yet another, more subliminal and latent neighbourly order, which is based on confidence drawn from the mere presence of a sort of generalized nearest other.

> I think it's got to do with urbanism, you know, that creates community . . . You are living on top of each other, so you can't be distant really. And yet again, and I might be contradicting myself, because, I found in a city, and that's something I like, you can be anonymous, you can be you know private, if you want to be . . . But, you know, there is a comfort in knowing people and not knowing people at the same time, trusting them and not trusting them, you kind of feel that if you had to trust them you could, but you are not gonna go out of your way to make you trust them, you're just gonna get on with it. And there is some kind of safety in city life as well, this safety in being surrounded by people, even though if you read the papers you would think people would rather walk away from you than help you. But I don't think that's particularly true. I think people would help you if you were in danger, I have a certain confidence in that.

What Ira describes here is a form of approved detachment, exercised between coexisting milieux on the basis of untested trust in the reliability of a 'generalized nearest other'. It is the mere co-presence of members of the metropolitan public that provides reassurance for her, for she is confident that in the event of 'real danger' this anonymous agglomeration of 'people' would turn out to be 'real neighbours', that they would not hesitate to engage in maintenance and care with regard to the milieu of the nearest in need. This does not necessarily have to be a big gesture like preventing someone from being mugged, but can express itself in much subtler ways.

We do not know where exactly Ira gained her deep seated trust, but perhaps she has shared the not so uncommon observation that on a winter's evening in one of London's parks a man walking his dog would point out to a jogger, 'take care, it's slippery down there'. Does this go along with the image of the Londoner as unhelpful, cold, and blasé towards the other (Ira: 'if you read the papers . . .')? What takes place in this brief encounter alludes to more than civil inattention between two people tactfully negotiating their personal spaces in public. The dog-walker's further going concern about the jogger's safety refers to their milieux as directed fields of actual and potential action. The dog-walker's concern is at least partly induced by the selfish yet also unselfish consideration of not wanting to have to attend to a jogger with a broken leg. Similarly, the jogger might now feel obliged to run cautiously, not only for his own safety but also in consideration of the dog-walker's interest. So, what we see in fact, then, is two people briefly concerned about maintaining the optional fields of their respective milieux.

This little example serves to illustrate the fine but important difference between Goffman's 'civil inattention' and Grathoff's 'approved detachment'. While Goffman's concept (1972: 385) stresses the avoidance aspects in the encounter of two people's *Umwelten*, Grathoff (1994: 37) stresses the presence of the nearest other as a positive contribution towards the maintenance of the other individual's milieux, for whatever selfish or unselfish reasons that might be the case. This difference comes clearly to the fore also in Ira's statement concerning the 'safety in being surrounded by people'. This implies that for her the generalized nearest other is not something to be avoided, but something that indirectly contributes towards the normality of big city everyday life (you're just gonna get on with it').

Finally, by the way in which Ira describes her relationship to the fellow Londoner, she clearly senses the inherent contradictions of approved detachment in practice. It is 'community' that allows 'anonymity', trust that does not want to be tested or exercised, closeness that requires distance, comfort from 'knowing people and not knowing them at the same time'. Her description goes right to the heart of what Kant asserts as the underlying antagonism of all forms of social organization – 'antisocial sociability' (1981: 208). Kant's term draws on the two contradicting forces in social life, that also, and in particular, underlie the individual's attempt to generate a milieu. On the one hand the individual's 'antisocial' will to order the world according to one's own interest, and on the other hand the 'sociable' necessity to relate to others and their interests in order to be able to pursue one's own interest to the best possible extent.

The 'distant' neighbour

The symbolic figure of the neighbour as the nearest other, involved in the maintenance of one's milieu, is not confined to describing relations between local

157

neighbours. Who qualifies as the nearest other is not necessarily a question of local vicinity, but first and foremost defined by the practical relevance or availability of the other with regard to care and maintenance work in the life-world (Grathoff 1994: 35). It could be argued that, in particular the extended milieu, that reaches beyond the actual 'here and now', is dependent on the maintenance work of the distant nearest other. That nearest other becomes a 'gatekeeper' with regard to fragments of the symbolic territory of the self, for example, distant posessions and obligations that are momentarily out of the individual's immediate reach.

Barbara, for example, relies on such a gatekeeper with regard to some of the important personal things she had to leave behind when going to the Philippines, and later London. Since with her financial budget frequent travel between London and the States is not really an option, Barbara relies on the maintenance work of this 'friend' in Nashville with regard to technical matters arising in her absence (such as car tax or health insurance).

> There is this friend of mine that I depend on a lot for helping me a lot before I left Nashville to get my business things in order. So, for instance, she helped me to get a safe-deposit box to keep some of the things that were important to me, to store there, and she has got one of the keys to that, so she has got access to it. Then she was willing to be co-signer and co-executor of my will, if something should happen to me while I'm in London.

This arrangement indicates a lot of confidence and trust in someone she is prepared to call a 'friend', yet this confidence has partly been eroded into serious concern with regard to the trustworthiness of that person as Barbara has not heard from her 'trustee' since leaving Nashville.

> And I've written to her twice and have not heard anything from her, and I'm wondering why, and that makes me suspicious, you know, whether she really was taking advantage of me.

This brings up the paradox of the 'distant nearest other', and with it a whole set of questions concerning trust across distance and the assumed priority of face-to-face interaction. With the extended milieu becoming a reflexively organized project across distance, it has by necessity to be 'opened out' to distant others, which spells vulnerability and risk-taking. Just as the extended milieu is no longer embedded in the stable social relations and moral standards of the local community and kinship networks, neither are the trust relations on which its situatedness is reliant on. As Giddens (1994: 121) points out, under these circumstances trust itself 'becomes a project to be "worked at" by the parties involved', and cannot be controlled by fixed normative codes'. Clearly Barbara faces a huge conflict in this regard. On the one hand she is clearly worried, as the other person does not act in a way that reassures Barbara's trust in her. On the other hand she is technically dependent on the good will of that 'friend'. Any open acknowledgement of mistrust would cut the lifeline to an important fragment of her extended milieu. Moreover, as no fixed rules have been established, on what grounds would she make her claim of mistrust? How often should the trustee be in contact, assuming that everything is 'normal' in Nashville and London?

What clearly does not help in Barbara's case is the fact that she belongs to the (relatively) disadvantaged when it comes to access to global means of transport and communication. Living on a low budget during her two years stay in London, she will fly to the States only once, and has forbidden herself the (expensive) use of the phone, thus mainly communicating via e-mail at work and via letters. With her trustee apparently not being linked up to e-mail, their 'trust project' across distance becomes unnecessarily complicated through lack of contactability, and is therefore perhaps more the exception than the rule.

In contrast, the 'settled cosmopolitan' Ulla's engagement in a 'transatlantic support service', maintained via global transport, can serve as illustration of the options given through globalized 'presence availability' (Giddens 1991: 122f.). Technically, global communications have compressed the world into virtual co-presence. And even if one is to associate co-presence with face-to-face interaction, global transport has certainly extended the number and distance of those that can come into consideration for practicable 'presence availability'. It is these circumstances, then, that allow an elderly lady in London to be placed on call for local maintenance work for friends in Newfoundland.

> So one year, for instance, I flew over [to Newfoundland] and looked after their home while they were having a holiday in Ireland, you see, that kind of thing.

Here we see the care-taker function for the localized elements of someone's milieu, which would 'normally' be associated with a local neighbour, temporarily handed over to an acquaintance who qualifies as the nearest other on the grounds of trustworthiness and presence availability. The necessary mutual trust for this transatlantic project developed, we have to recall, through repeated co-presence, namely, regular visits across the Atlantic. Particularly important in the binding experience was the shared grieving over a person that was sister and friend respectively, thus generating for this support service across distance an additional embeddedness in affective like-mindedness.

This tendency to lift relationships of care and maintenance with regard to someone's milieu out of the relationship between local neighbours, seems to apply not only to the milieu that extends across distance, but can also be found in the organization of milieux within the locality itself. A subtle shift in this direction can be detected from the account on 'neighbourly' relations given by Sarah, the 'mobile cosmopolitan'. She describes the relationship to her immediate neighbour on one side as one where 'we don't really socialize, but we chat over the garden fence and things', while the neighbour on the other side she describes by saying 'I don't really know at all, I'm not even really sure if anyone is living there at the moment'. As someone who is travelling a lot, Sarah will have to rely on that former neighbour's function as an invisible 'gatekeeper' regarding the local boundaries of her milieu. However, the actual care-taker function with regard to maintaining her milieu in her absence lies with a 'neighbour' down the road.

> A pair of keys is with my neighbour a few doors down the road, because he is a builder, and he does often bits of work on my house, and, you know, I can trust him to come and go as he pleases. So he has the keys, so that he can go in while I'm not here.

What qualifies this builder down the road to be the nearest other, symbolized through the handing over of keys, is not so much local vicinity, though 'presence availability' is important in this case too. The point, however, is that this 'other' becomes a 'relevant other' to Sarah's mobile and extended milieu through a combination of personal trust and expertise. Not every builder would gain Sarah's trust to come and go as he pleases. And not every person that Sarah would trust with unconditioned access into her milieu would also have the necessary professional maintenance skills. For a busy professional and traveller like Sarah, this is the perfect arrangement with regard to the local aspects of her milieu. To fully appreciate this seemingly trivial arrangement, we have to recall here that we do not talk about property maintenance in the narrow sense here. Sarah's house is the core of what earlier on, following Goffman, was described as symbolic territory of the self. It is 'my base', as she pointed out. This is where feelings of 'belonging' and yet also 'independence' are centred for Sarah. Knowing that things are 'in order' in this place, gives her the 'free play' to engage in entertainment across London and travel across Europe and the world. In this regard the 'builder down the road' is the nearest other, with whom Sarah is engaged in a 'trust project' in the above-described sense. He is 'on call' with regard to domestic items and arrangements crucial to Sarah's wider milieu (Goffman 1972: 51f., 337f.). Like Barbara, Sarah has had to 'open out' her milieu to the relevant other, in order to maintain a mobile and extended milieu. Unlike Barbara, however, Sarah has ongoing evidence on the spot for this being a working arrangement with someone in 'presence availability', rather than having to have trust a distant other who is almost out of reach.

The disembedding of trust and care relationships involved in the maintenance of

160

a milieu finds yet another expression in Rolf's case. Rolf, who lives in a purpose-built block of flats, says that he would rather face the risk of being locked out than place a key for emergencies with neighbours. If such a case should arise he would rather place his reliance on the services of a key cutter or wait until the landlady comes over from Wales. What Rolf is effectively saying is, that instead of going through the laborious work of generating a trust relationship with one of his neighbours, he would rather rely on (yet) anonymous 'experts', if one is, for the sake of argument, prepared to see the key cutter and the landlord as professional roles 'which have validity independent of the clients who make use of them' (Giddens 1993: 18; also 1994: 27f.).

Again, we have to understand Rolf's decision in the specific context of his milieu. Having just separated from a long-term girlfriend, and faced up to a long-suppressed aspect of his 'self', Rolf is in the process of a reflexive reorganization of his milieu. Moving into another area of London and into a flat of his own is part of that process. There is an affinity to this place, which is expressed when he says 'I love this place, this is where I have regained my freedom'. Accordingly, he is not prepared, at the moment at least, to jeopardize this newly gained 'free play' in his milieu.

> I mean, I don't mind [not having contact with his neighbours] . . . , I have my freedom here, and I don't want to be tied into some kind of obligations with some neighbour, like you have to invite them because they had you over for a glass of wine or something, I don't need that sort of stress at the moment.

As Giddens (1994: 121f.) points out in this respect, the reflexive restructuring of the self is a risk-taking exercise that includes the 'opening out' of the individual's milieu. Consequently, the individual has to take into account all available options to secure this project, including those provided by 'expert systems' in the broadest possible sense. In this light, Rolf sees the involvement with local neighbours clearly as too big a risk to take, and the risk of having to 'wait until the landlady comes over from Wales' in case of an emergency, as the lesser evil. What we are left with in Rolf's case is an extreme illustration of disembedding and individualization as complementary processes (Beck 1997: 94ff.). Once seriously engaged in the process of conducting and arranging a 'life of his own', Rolf finds himself, more out of choice than necessity, without any nearest others.

Neighbourhood community turned socioscape

Positioned at the core of a landscape of global sociocultural flows, the socioscape of the global city itself becomes diversified but also more transient. Two-fifths of London's population are 'Londoners by adoption'. And within its population of about

6.4 million there is an annual turnover of approximately 350,000 people, with one-third of newcomers arriving from outside the UK (Hall 1990: 7, 13). Yet even for those who call London their home, for however long that might be, it does not mean that they stay put. They might socialize across London (Ira), spend their weekends sailing in the South of France or attending singing workshops in Venice (Sarah), they might spend the week doing business somewhere in Europe (Harold), fill a gap between two posts in Asia (Barbara), have just come back from a holiday in native Cyprus (Nicos), be looking after someone else's house in Newfoundland (Ulla), might just have moved house after a complete change of life style (Rolf), or might have moved out of the metropolis altogether because they could not find peace of mind in this transitory environment (Herbert).

Indeed, Herbert earlier remarked on this transitoriness in his life-world as one of the reasons to leave the London he loved so much, and to resettle in a village community.

> Because, when I was young, people lived in these streets for generations. But now the population in London is so transient, that they move in and they move out again, you know, there is not the same association with a particular area and the same road any more.

One might feel tempted to put this remark into perspective by ascribing it to an old man's nostalgia for the London of 'those days'. But the transitoriness of local life is something equally remarkable for someone who is actually part of this transient, relatively unattached life style. Rolf gives a vivid account on his first impression regarding the frequent turnover of neighbours, the little mutual involvement, and the anonymity he felt at times.

> In the first autumn when we arrived, they lived already in that other flat. Both are some kind of artists or actors, but don't ask me at which theatre, you would just bump into them occasionally, do the typically English small talk, like 'How are you? Oh, I'm fine, how are you? Nice weather today, isn't it?'. We only met properly once, when they felt they had to invite us to William's garden party, it was his birthday, I think. But then they moved out in August, and the flat stood empty, until I moved out as well [when separating from his girlfriend Petra]. Then another couple moved in, and by Christmas, Petra and I drove back to Germany together, they hadn't even realized that she would not be there over Christmas. So, I think the whole thing is somehow more anonymous in such a big city.

Rolf seems to rationalize this experience for himself as a question of size. But, as we saw in the previous section, Rolf later developed an outlook on life in the world

162

city that makes detachment from neighbours an essential aspect of his experimenting with the 'freedom' of a new life style. Perhaps Herbert is closer to the heart of the matter when he refers to it as a problem of 'attitude', of 'lack of get together spirit'. Sarah remarked casually with regard to her involvement with neighbours or the local community, that she is 'just never there when one meets the next'. And Ira told us that she would much rather be on her own when she can not afford a travelcard to meet friends somewhere else in London. Harold, in turn, talks about his neighbourhood affectionately as 'the village', but in fact entertains what amounts to no more than a weekend relationship with that village.

There is good reason to assume that, with people increasingly able to conduct their lives in what was earlier described as mobile and extended milieux, local social relations will be transformed as well. Simplifying, one could say that the attachment of people's milieux to non-localized sociospheres allows them, perhaps even forces them, to be more detached from local neighbours, and less involved in local social relations. Also, as pointed out previously, for the extended milieu the nearest other is not necessarily the next door neighbour. The more the distant nearest other is relied on for maintenance work concerning the extended territory of the self, the more the local neighbour becomes less relevant, in both the common sense and the theoretical meaning of the word. The neighbour disappears, literally, as in Rolf's account on the turnover of people in his house, and metaphorically, in terms of mutually approved non-involvement, as well as a symbol for the 'decline of community [spirit]', as declared by Herbert.

But has it ever been different? To appreciate the transformation that has taken place, we will have to contrast the accounts given by Rolf and Herbert with a picture of the London street community of the past. Herbert gave us a glimpse of that past earlier, when he described the street village communities in South London's Elephant and Castle area as he experienced them before and during the Second World War.

> In these working class areas, every street or every two streets were like a village . . . you lived together, you knew each other's business to an extent. And if anyone died, someone would go around for a collection for flowers, and the blinds would be drawn when there was a funeral.

This picture of a close-knit street village community, based on personal acquaintance, and symbolized by shared local custom, is underlined in Young and Willmott's classic *Family and Kinship in East London* (1962). In Chamberlain's narrative-based study *Growing up in Lambeth* (1989), he paints a picture that complements Herbert's account of the existence of a street village community:

> Neighbours knew one another's business. Family noises and rows filtered through into the street. Within the neighbourhood of the house, how

people slept, when they slept, where they slept, what work was done and taken in, which room was vacant, who was dirty, who was clean, who was healthy, who was sick, who was in work, who was without . . . all of this was *common knowledge* [my emphasis].

(1989: 14)

These neighbourhood communities were galvanized by almost exclusively localized social relations, perpetuated by shared conduct of daily practices such as washing, mangling and cleaning, and controlled by the strict 'standards' that were necessary for the use of shared facilities like wash-house and toilet.

The points of contact and sharing were too numerous to allow for harmony all the time. Who last cleaned the stairs, used the lavatory, used the wash-house: all were points of potential friction . . . Under such conditions it was necessary to generate ground-rules for survival, and it was these *ground-rules* that became the heart of a neighbourhood [my emphasis].

(ibid.: 15)

These 'ground-rules' would cover areas of everyday life 'as fundamental as trust' and 'as superficial as cleanliness and noise' (ibid.: 17). Binding for everybody, they provided the means to maintain the 'closed society' of a local neighbourhood community, with a shared local culture, exercising mutual care and intimacy between its members on the one hand, and punishing and discriminating the outsider (ibid.: 18).

The difference between these relationships of local care and intimacy on the one hand, and, on the other hand, the detached attitude between neighbours that Rolf and Sarah described, is pinned down rather nicely in Bailey's (1995: XV) snapshot of neighbourly life in South London,

Talking in the street was not a sign of being neighbourly – that was mere politeness. True neighbours sat in your chairs, and shared their joys and sorrows.

What we have to bear in mind at this point is that the portrait of closeness and intimacy in the previous narrative sequences reflects a shared local culture that developed more out of necessity than choice. As Pahl (1973: 103ff.) makes clear, when 'for good or ill, a given collectivity will surround one for the rest of one's life, one is obliged to have a stake in the local situation'. To call this type of local culture 'community of common deprivation' is apt, as the option to maintain alternative – non-local – social relations were not given. As Chamberlain (1989: 14) points out in this regard, 'A family which moves 2 miles away is completely lost to view. They never write, and there is no time and no money for visiting. Neighbours forget them'.

In other words, the neighbourhood communities described by Herbert, Chamberlain and Bailey, were confined to what could now be described as a single local 'sociosphere' (Albrow 1997: 51). They had no alternative than to participate in a locally confined 'field of concern and relevance'. Or so it seems. And yet, another local voice mentioned in Bailey's study (1995: XIV), states something remarkable for the South London of the generation before Herbert.

> During our fifteen years in the one house we never had the slightest acquaintance with our 'semi-detached', nor with the people round, although we knew several by sight and gave them nicknames.

Bailey accounts for this, at that time, exception to the rule, as follows:

> Mr Vivian, his wife, sons, and daughter were consequently a 'cut above' people such as my parents, who were Londoners with lots and lots of neighbours.

It is certainly no overinterpretation to see the metaphor 'a cut above' as an early indication for a stratification of sociospheres. The Vivians will have had their own networks of social relations with nearest others, those neighbours of like-mindedness, somewhere across the metropolis. They and their local neighbours lived in different fields of concern and interest, lived in different social worlds with different degrees of mobility, and filled with different social practices. For those living their everyday lives 'a cut above', there was no need to get deeply involved in localized social relations, while those deprived of the alternative to local life had no choice but to have lots of neighbours.

This anecdote indicates a transition of local social life towards the state of affairs as described earlier on by Rolf, Sarah and Herbert for today's London. The local neighbour ceases to be the nearest other ('not the slightest acquaintance') in terms of contributing to the maintenance of a respective milieu or sociosphere. We see early indications of approved detachment here. There is non-involvement in the relationship between the Vivians and the 'rest' of the street, but at the same time a high degree of mutual awareness, as shown in 'we . . . gave them nicknames', normally based on someone's peculiar features or habits. Maybe the Vivians had nicknames for the 'rest' of the street. This constellation might have been repeated elsewhere in London. It certainly resembles the 'Thrale and Johnson' setting and their non-involvement in Streatham's local community, as described in chapter 5.

In the 'transitory' London described at the beginning of this section by Herbert and Rolf respectively, the Vivians' experience would not be the privileged exception, but the norm. These early indications of a transformation in local social relations indicates the long-term process of a differentiation of local community and culture into a socioscape – a loose form of coexistence between different milieux and their respective sociospheres. It would be wrong to ascribe this differentiation process entirely to processes of globalization. However, processes of disembedding in form of the extension of the individual's milieu, and their increasing attachment to non-local sociospheres, has certainly contributed to the transformation of local community into a socioscape of largely unrelated personal and social worlds. Local community without shared local culture becomes locality, a place where different milieux and their sociospheres coexist, perhaps intersect, and open for being interpreted and imagined in various and contradicting ways. As Albrow (1997: 52) aptly points out with regard to research conducted on the socioscape in the London locality Tooting:

> For each person their place in the locality represents a point where their sociosphere literally touches the earth. But for each person who is viewing other people there can only be a very partial idea of the relevance of the locality for other's sociospheres. What they experience is not, therefore, in general anything like the traditional concept of community based on shared local culture. Rather they engage in something like a cavalcade where passing actors find minimal levels of tolerable co-existence with varying intimations of the scope of other people's lives.

This is not to say that everyone in London lives in a locality turned socioscape as outlined in an ideal typical picture in the above quote. This transformation is gradually, perhaps surfacing more clearly in Inner London areas such as Tooting than in the suburbs of Greater London.

In any case, these emerging socioscapes are detectable through the biographical accounts of the milieux participating in them. Sarah and Harold are clear cases of milieux literally 'touching down' in a locality. Sarah, as mentioned earlier, sees her home in Peckham in the first instance as a 'convenient base for travel', and 'easy access to the City', her place of work. Even more extreme in expressing his view of the locality being first of all a 'base' from which to be mobile is Harold, the 'global business man'.

> Geographically, where we are located, it's easy to get out of London, because we are very close to the North Circular, we are not too far away from the M1, we are not too far to get to the M25, going east and west is very easy for us, OK . . . We live in a good enough place to travel, we can get to any place from where we are, it's easy to go to Dover and take a ferry, in an hour and a half we can be in Dover . . . We located so that we

were near one of the main underground lines, OK, so that our children could go directly from where we live to the American School without having to change trains, which is nice.

Harold's account gives evidence of a locality clearly chosen on the basis of it being a place that technically fits in the structure of relevances of his milieu, taking into account the needs of global and metropolitan mobility. But for the less mobile, too, the locality seems to be a collage of local spots that fit the relevances of their milieux. For Ira, we can recall, 'her' Brixton was mainly the Brixton of the underground station entrance, with all the cosmopolitan hype around it, and her residential street. And for Barbara the picture looks similar. Knowing that she will live in Streatham only for the next two years, and with her social life evolving around Filipino London, there is no attempt from her side to become involved with the local community. As far as Streatham is her 'home', it takes again the form of a collage of relevant locales.

Well, I feel at home here, in this house. And I feel at home knowing where some things are in Streatham. But I don't really feel like Streatham is a place I know very well, other than the Commons (were she takes her evening walks), the house, and a few stores, like the 'Three Bakers', but I wouldn't call this community 'home'.

In the light of these accounts it becomes clear that locality is not just a neutral, empty space in which to 'place' different milieux, but the locality itself, viewed through the filter or perspective of the milieu, appears as a landscape of shifting relevances. Following Schütz (1971: 93), one could say that the locality shows 'hypsographical contour lines of relevance . . . which in turn indicate the distribution of interests of an individual at any given moment with respect both to their intensity and to their scope . . .'.

A similar point could be made with regard to the 'others' who co-exist in the locality. With the perspectival view on the locality and its inhabitants further enhanced by the different time–space patterns of their everyday lives, people belonging to different sociospheres might rarely cross paths, or indeed fail to 'see' each other. For example, we can recall Sarah's earlier statement concerning her patterns of work and socializing:

But I found that, particularly in winter months, I'm just never there in daylight when one meets the next, I'm always here in the hours of darkness, so you just don't really see people, and you don't see changes happening.

Sarah, in this regard, sees her neighbourhood consisting of two basic sociospheres, that of the retired elderly, and that of the busy professionals.

I mean, you just don't get people sort of sitting at their windows and watching what's going on, you find the people who really do know what's going on are the retired elderly people who do sit at their windows and look and see what's happening in the street.

So, is this then a case of the new version of 'absentee landlordship' (Bauman 1998: 3), where the global elite enjoys increasing mobility and independence from local issues at the cost of those who are left to look after the neighbourhood? Perhaps. But it is also possible that Sarah might find that some of these elderly retired people who seem to sit all day long at their windows, operate in a similar time–space pattern as Ulla does, that is, confined to the neighbourhood for most of the year, but then off to Newfoundland or Sweden for a couple of months.

As far as the actual socializing in non-local sociospheres is concerned, it is not just Sarah and her City friends who socialize in what she called 'third territory', outside their respective neighbourhoods in place and venues across London that are convenient for everyone involved. Barbara, on a very low budget, hardly spends any spare time in Streatham, but socializes with other American expatriates, and manages to visit the London museums with her Filipino friends. And Rolf only comes home to change clothes before going off to a choir rehearsal or to enjoy London's West End gay scene. In fact, in each of the milieux of our eight Londoners could we find indications of their involvement in sociospheres that extend beyond the locality across metropolitan space, or indeed global space. The only exception, perhaps, is Herbert, who managed to regain an almost closed local sociosphere by moving into a village community.

The transformation of local community into a locality with a shifting socioscape does not mean the end of local social contact or neighbourly involvement. As shown earlier, the local coexistence between milieux has its own social order(s). In this context we introduced the notion of approved detachment as a general form of engagement between neighbouring milieux in order to show acknowledgement and appreciation of the presence of the nearest other. This was illustrated by the symbolic exchange of cutting shrubs and a bottle of wine in return, in which Harold and his neighbour engaged. This exchange indicates a form of social engagement that is independent of face-to-face interaction, thus perfectly practicable for milieux whose time–space patterns and sociospheres hardly intersect.

If milieux touch or intersect in public, however, we can expect their encounters to be guided by forms of 'civil inattention' (Goffman 1966: 83ff.) and/or 'sociability' (Simmel 1969: 40ff.), depending on the kind and setting of the encounter, and on the degree of mutual involvement required respectively. With neighbours in a residential street often being no more related than regular passers-by on the High Street, the same rules of acknowledgement and indication of interest in non-interference apply. Ira had such a 'nodding relationship' with her neighbours.

'Sociability', the art of the sociable moment, on the other hand is more relevant to encounters in settings such as the local newsagent or the local playground.

And finally, there might even be some sort of directed action, when the overlap of interest is sufficiently 'galvanized' around a certain issue. As Sarah mentions in this regard, 'It always needs some big issue to make everybody aware of something they don't want to happen and to make them bump together'. While in many cases this 'big issue' might be to fight the planning permission of an unloved neighbour, this galvanizing process can also evolve around issues of greater practical and symbolic significance, as shown in chapter 5 with regard to Streatham and its department store. But, we must not forget that these involvements are a question of local 'relevance' in the above defined sense. For the milieux involved it does not mean the resurgence of an all embracing local community.

However, with regard to the different forms of local social contact, it could now be claimed that sociability is perhaps not a possible form, but *the* form of social interaction that is most appropriate for a local setting of coexisting milieux and their sociospheres. Simmel (1969: 45f.) describes 'sociability' as a 'play form of sociation' that does not rely on any pretext such as wealth or social position, has no objective purpose, and is entirely dependent on the personalities of those involved. Sociability generates an 'as if' world that has no real consequences for those involved, as it aims at 'nothing but the success of the sociable moment and, at most, a memory of it'. There are two aspects now that distinguish sociability with regard to its potential for social interaction between milieux belonging to different sociospheres. It allows people who otherwise belong to different social worlds, 'to become sociable equals' for the casual moment (ibid.: 49). More importantly, as pleasurable as it might be, sociability as a form of social interaction is an empty play, 'lifted out of the flux of life' (ibid.: 43). It thus provides the appropriate social form for a successful moment of pleasurable interaction between people who otherwise live in different social worlds. Sociability allows them to detach themselves from the social gathering the next moment without any social consequences. Harold recalls one such particular moment of sociability, the memory of which is still with him today:

> I think moving into that village was nice. We met nice people, and they were very friendly. When we moved in they brought over coffee and biscuits. We were a phenomenon, you know, having Americans moving into their place . . . So in terms of the village, the village is nice, the village is friendly, no problems there.

Harold was never expected to return this favour in terms of inviting his neighbours round for coffee, or by bringing them presents from his European travels. What did count, for both parties involved, was the success of the sociable moment in an ideal

169

world 'in which the pleasure of the individual is closely tied up with the pleasure of the others' (ibid.: 48).

It is now interesting to note that this, perhaps repeated, enacting of sociability makes Harold talk about his neighbourhood as a 'friendly village'. That he feels as if living in an 'as-if' community does not contradict the assumption of London localities being socioscapes of some kind. Sociability allows him to view the locality, which before he described as 'a good enough place to travel', as the perfect shelter after a week travelling in Europe ('the village is friendly, no problems there'). Harold is in his 'friendly village', no more involved than Sarah in her neighbourhood, or Ira in her residential street. And indeed, looking a bit closer into Harold's 'village', it turns out that he is not the only one who 'sticks out' from what would otherwise pass as a homogenous suburban street village. He is aware of the coexistence of at least two other sociospheres. There are on the one hand 'a few Pakistani or Indians' and, according to him, there is a 'silent agreement not to sell [property] to them'. And then there are those whom he describes as 'newly rich yuppies', with their own ways of doing things and their own time–space pattern.

> The level of community togetherness has dropped with these people, because they went to work very early, and came home very late. On Saturdays they went to, you know, whatever activities they are involved in. And they had the money to be able to afford to hire somebody to cut their lawn and to do the fixing up, but a lot of people in the village couldn't afford to have somebody to come and paint their houses for instance, they had to do it themselves. So you had this separation too.

If the fact that Harold, who himself is never there during the week, still considers himself as part of the village community, and points towards 'others' with different time–space habits and involved in obscure Saturday activities, shows anything, then it is the deeply perspectival character of a neighbourhood as a socioscape. From the perspective of different milieux and through the formative power of the sociability games within different sociospheres, the locality and its inhabitants can look very different indeed.

Sarah gives another vivid example of this ongoing process of construction. Though she hardly knows her neighbours, she has a clear idea of what her neighbourhood is actually like. It consists of 'lots of different nationalities' and 'people like me, that is middle class people with quite good jobs and very good education'. This assessment is a projection of her milieu and the 'standards' of her sociosphere – cosmopolitan values, good education, good income – onto anyone else in the neighbourhood. She also convinces herself that Peckham 'is friendlier than a lot of other areas in London'. Still, Sarah knows at the same time that this picture of her neighbourhood is not shared by everyone. Certainly the local newspapers tell a slightly different story of increased crime and violence.

> Maybe it's burying my head in the sand a bit, but I just thought 'No, I don't want to read about things like that'. You know, I just take reasonable precautions to avoid them happening to me, and I don't really want to know about them otherwise.

What this statement finally reveals is someone carefully navigating the sociospheres not only of London, but also her own neighbourhood. The argument developed in this section thus makes Appadurai's (1992: 297) assumption, that the shifting landscape of global culture is somewhat anchored in 'relatively stable communities', questionable at least with regard to London's socioscape. The picture drawn in this section suggests that not only is London as a whole a fragmented cosmion, a landscape of different social and cultural worlds, but that its neighbourhoods, at least for the people who live in it, resemble local socioscapes more than local communities.

CONCLUSION: THE END OF
THE WORLD CITY?

People's milieux are affected by processes of disembedding and re-embedding, just as much as the world around them. Formerly stable reference points, like the local community, are increasingly ruptured. The biography becomes an open and ongoing project that has to be maintained across distance. As the world seems to shrink, our personal worlds tend to expand. Time–space distanciation surfaces in our milieux as extension of the 'symbolic territory of the self'. We relate to significant places and others on a daily basis. Time–space compression in turn is experienced as globalized presence availability and as a convergence between 'here' and 'there'.

The lived reality of the extended milieu, as it has surfaced in various ways in the narrative accounts throughout the book, can be taken as evidence for the individual's active effort to generate and maintain her or his world of everyday life within a global culture that emerges within and outside the globalized world city. This is not the same as saying that living in extended milieux is entirely the outcome of globalization processes. Neither is it a natural outcome of living the globalized world city. All that is maintained here is that people in today's world are affected by, and actively participate in, various aspects of diverse globalization processes. With a place like London at the core of a new global cultural economy, this should apply to those 'Living the Global City' (cf. Eade 1997) more so than it does to others.

The lived reality of the extended milieu is evidence for processes of disembedding and time–space compression at work in people's life-world. The extended milieu can be seen as a symbolic marker of transformations in wider society. As such it seems to symbolize two such transformations; on the one hand, the transition from modern to global society, and relatedly, the transformation of the modern metropolis into a globalized world city.

Modern mobility, ideally or typically speaking, equalled dis-placement and dis-placed milieux. It was a reflection on the radical uprooting and the subsequent mobilization of milieux in the context of modern industrialization and urbanization, and led Berger et al. (1974) to speak of the 'homeless mind' as the symbolization of a spatial, but even more so, metaphysical restlessness. Cut loose from its familiar environment, according to Berger et al., the uprooted milieu would never truly regain 'home', despite moving through a succession of places and social worlds.

On the one hand, globalization processes can be seen as continuing these uprooting tendencies, and thus might generate 'homeless minds', such as Herbert, moving on several times in the attempt to regain a local milieu. On the other hand, however, globalization processes open up new forms of re-embedding that allow, for example, Barbara to feel 'at home' in Filipino London. Both, the access to global communications and the active participation in the emerging landscape of global sociocultural flows, allow for life in-between places. The extended milieu inhabits space rather than place, be that the space which stretches between Brixton and Kingston Town, or that links Manila and Filipino London. However, life across distance and in-between places does not sit equally comfortable with everyone. It is not necessarily the 'good life', as is evident, for example, from Costa's description of being torn between London and Cyprus. In this regard the extended milieu reflects the ambiguity inherent in the process of globalization itself.

Globalization does not draw modernity simply to a close, but opens up the human project again. It allows for new forms of social interaction, and encourages new forms of social and individual identity just as much as it challenges established social forms. The extended milieux is in this respect one indicator for the opening out of the biography and the transformation of the individual's milieu into a reflexive project.

Just as global culture and global society are not simply a consequence of modernity, so is a globalized world city not a simple culmination of metropolitan development, or a modern world city writ large. Again, there is continuity and change involved here. The globalized world city is a metropolis of sorts in that it is at the heart of the global cultural economy, and subsequently at the heart of a landscape of flows of people, cultures and symbols. However, it is not necessarily the 'melting pot of races, people and cultures' that requires social integration or/and spatial segregation (Wirth 1969: 69f.). While time–space compression and other globalization processes might intensify the flows of ethnic and cultural diversity that are touching down in the globalized world city, they also change the ways in which this diversity is arranged within the global city's cosmion. The globalized world city, one could argue, provides a global cultural economy *en miniature*. It consist of a landscape of comparatively unrelated and non-localized social and personal worlds. The possibility of extended milieux and their attachment to non-localized sociospheres implies a new 'free play' between locally coexisting worlds. Apart from the descriptive accounts of the 'potpourri' or the 'salad bowl', Albrow's notion of socioscape is possibly best equipped to grasp this constellation of local coexistence based on mutual non-involvement.

These new forms of approved detachment within the cosmion of the global city are of an ambiguous nature as far as their cultural consequences are concerned. They allow for both tolerant engagement with the strange, the other, and the novel, but also complete ignorance and detachment. The globalized world city is home to both, the metropolitan connoisseur like Ira, and the global philistine like Harold, who is at home in airports and hotels all over the world, but avoids the cosmopolitan variety at his doorstep. In the light of this ambivalence it seems difficult to maintain Mumford's optimistic view with regard to the cultural function of the world city. For Mumford (1991: 649, 653), the world city has to serve 'the progressive unification of mankind itself'. It has to serve 'as an essential organ for expressing and actualizing the new human personality – that of "One World Man"'. This humanistic optimism is difficult to share, in particular because it is based on the assumption that the world city will hand on to the smallest urban unit the cultural resources that make for world unity and cooperation (ibid.: 639). Instead, what can be detected from the narrative accounts of life in the smallest urban units of the world city, its neighbourhoods, is that these, too, resemble socioscapes, consisting of relatively unrelated social and personal worlds. Mumford's account assumes the world city to be an epitomization of a single world society. But it seems equally plausible to see the world city's cosmion as a prism that reflects and refracts the non-isomorphic, deterritorialized, overlapping, and disjunctive order of the landscape of global sociocultural flows.

The extended milieu is an indication of this new glocal landscape. It consists of glocalized zones of familiarity, practical competence, and symbolic situatedness, maintained across distance via individual mobility, access to global communications, and globalized presence availability of the nearest other.

NOTES

Chapter 1

1 Referring to Hannerz (1980), to mention just one of the more intriguing amongst the vast number of studies having 'urban' or 'city' flagged in their titles.
2 The term should not be confused with Robertson's notion of 'miniglobalization', which refers to the fact that 'historic empire formation involved the unification of previously sequestered territories and social entities' (1993: 54).

Chapter 2

1 In view of Crook's recent (1998) criticism of the concept of everyday life as a 'minotaur' with undisclosed ontological and epistemological baggage, it seems necessary to clarify at this point: 'Phenomenology of Everyday Life' in the way I set out to use it as a dimension of a 'Phenomenology of Globalization', initially implies nothing more than the reconstruction of empirical social worlds through biographical reference points. Subsequently then this 'material' is being used for an empirically grounded demonstration of how individual actors actively generate lived space and place *within* global society. It is exactly the intention of this book to show how in different ways people participate in the concrete structuration of a global life-world, not however, to defend the 'purity' of everyday life against the onslaught of globalization processes, perceived as 'external' to everyday life.
2 Accordingly, I will leave aside the admittedly interesting, but at this point secondary, philosophical issues as to whether Cassirer and Goffman can be justifiably claimed for the Phenomenological approach.

Chapter 4

1 The term is used here in a different way than in Hannerz's argument (1992: 241). While Hannerz describes the anti-cosmopolitan global traveller, not willing to engage with local culture, Herbert is the anti-cosmopolitan mobile local, unwilling to engage with his microglobalized social environment in order to protect his local milieu.

175

Chapter 5

1 The figures refer to internal statistics of 'St Anne and St Agnes' church office as of January 1995. I am grateful to Pastor Englund and Ms McGee for allowing me access to these figures.
2 St Anne and St Agnes is in fact held on lease from the Dean of St Paul's Cathedral (Willows 1988: 179).
3 This is expressed even more clearly in the German translation, where it says that the 'established / outsider' configuration was generated between different groups of people 'die sich allein durch ihre Wohndauer am Platz unterschieden' (Elias and Scotson 1993: 11).

Chapter 6

1 Name altered.
2 Name altered.

Chapter 7

1 The notion of 'intuitive space' attempts to grasp the move from perceptual seeing towards imaginative seeing. The German term 'Anschaungsraum' appears to stand better in the context of Cassirer's explanation of 'symbolic space', as 'Anschaung' plays on the notions of both (sinnliches) 'An-sehen' and (geistiges) 'Vorstellen'.
2 The deeper meaning of Cassirer's intended play on words is again partly lost in translation, as 'erfassen' (grasping) in German can mean both, something physically 'er-greifen' and/or conceptually or intuitively 'er-fassen' (i.e. deuten, begreifen).
3 It is consistent with his argument on the intensification of stimulation that Simmel does not make this differentiation. He used the the term milieu as equivalent to environment. This is coherent with his assumption of an unmediated relationship between individual (and its nervous system) and (metropolitan) environment.

Chapter 8

1 As Srubar (1988: 9ff.) in his elaborate account of Schütz's concept of the life-world as a cosmion points out, the term is not coined by Schütz, but it was he who was devoted to analyse the role of symbols in the process of the internal structuring of the world of everyday life. In this regard the term best reflects Schütz's continuous attempt to understand the life-world as a symbolic order.
2 I want to establish at this point a clarification concerning the use of the 'symbol' and 'symbolic reference'. Schütz, in some of his argument, differentiates between 'sign' and 'symbol', reserving the term 'symbol' for links between the world of everyday life and other 'provinces of meaning' (i.e. religion, etc.). I shall use the term 'symbol', rather unconcerned for any appresentational reference that transcends the actual 'here and now', i.e. links 'here' and 'there'.
3 The phrase is borrowed from de Certeau (1988: 91) but slightly altered in its meaning. While de Certeau describes the erosion of the city's fixed structures and spatial forms through the subversive and creative power of urban practices, I simply imply in

this term the outwards-looking tendency of the metropolitan cultural environment as it constantly refers to places and happenings elsewhere on the globe.

4   With regard to future research it is interesting to note here that the affinity between the global *landscape* of cultural flows, as used by Appadurai, and the notion of an extended milieu as based on Schütz's concept of relevances that stretch across time–space, is underlined with regard to use of terminology. Schütz explicitly refers to the cartographical notions of 'isohypses' and 'hysographical contour lines of relevance' in order to grasp the links between 'here' and 'there' as a problem of intensity and overlapping patterns, rather than plain distribution (1971: 93).

5   The reader will notice that Schütz's notion of 'multiple realities' (1967: 340) has inspired this idea of a 'multiplicity of cosmions'. But again, similar to what has been said under note 2, while Schütz's 'multiple realities' or 'provinces of meaning' are referring to different cognitive styles that constitute realities *besides* the reality of everyday life, 'multiplicity of cosmions' attempts to grasp the coexistence of different frames of meaning *in* the world of everyday life. Srubar (1988: 251) sees the idea of a plurality of 'cosmions' already indicated in Schütz.

6   Also *Vermöglichung*.

BIBLIOGRAPHY

Albrow, M. (1993) 'Globalization', in W. Outhwaite and T. Bottomore (eds) *The Blackwell Dictionary of Twentieth Century Thought*. Cambridge, MA: Blackwell.

Albrow, M. (1996) *The Global Age: State and Society beyond Modernity*. Cambridge: Polity.

Albrow, M. (1997) 'Travelling beyond local cultures: socioscapes in a global city', in J. Eade (ed.) *Living the Global City: Globalization as Local Process*. London: Routledge.

Albrow, M., Eade, J., Wasbourne, N. and Dürrschmidt, J. (1994) 'The impact of globalization on sociological concepts: community, culture and milieu'. *Innovation*, 7(4): 371–89.

Appadurai, A. (1992) 'Disjuncture and difference in the global cultural economy', in M. Featherstone (ed.) *Global Culture: Nationalism, Globalization and Modernity*. London: Sage.

Aristotle (1968) *The Politics* (ed. R. McKeon; trans. B. Jowett). New York: Random House.

Arnold, F. (1886) *The History of Streatham*. London: Elliot Stock.

Augé, M. (1995) *Non-Places: Introduction to an Anthropology of Supermodernity*. London: Verso.

Bahrdt, H.P. (1974) Umwelterfahrung: Soziologische Betrachtungen über den Beitrag des Subjekts zur Konstitution von Umwelt. Darmstadt: Wissenschaftliche Buchgesellschaft.

Bailey, P. (ed.) (1995) *The Oxford Book of London*. Oxford: Oxford University Press.

Bauman, Z. (1998) *Globalization: The Human Consequences*. Cambridge: Polity.

Beck, U. (1992) *Risk Society: Towards a New Modernity*. London: Sage.

Beck, U. (1997) *The Reinvention of Politics: Rethinking Modernity in the Global Social Order*. Cambridge: Polity.

Berger, P.L., Berger, B. and Kellner, H. (1974) *The Homeless Mind: Modernization and Consciousness*. Harmondsworth: Penguin.

Borer, M.C. (1977) *The City of London: A History*. London: Constable.

Brandon, P. (1977) *A History of Surrey*. London: Phillimore.

Brenner, N. (1998) 'Global cities, glocal states: global city formation and state territorial restructuring in contemporary Europe'. *Review of International Political Economy*, 5(1): 1–37.

Bromhead, H.W. (1936) *Streatham's Beginnings*. London: Streatham Ratepayers' Association & Streatham Antiquarian and Natural History Society.

Budd, L. and Whimster, S. (1992) (eds) *Global Finance & Urban Living: A Study of Metropolitan Change*. London: Routledge.

Cadman, D. and Payne, G. (1990) *The Living City: Towards a Sustainable Future*. London: Routledge.

Cassirer, E. (1953) *The Philosophy of Symbolic Forms*. Vol. I 'Language'. New Haven and London: Yale University Press.

Cassirer, E. (1957) *The Philosophy of Symbolic Forms*. Vol. III 'The Pheneomenology of Knowledge'. New Haven and London: Yale University Press.

Cassirer, E. (1970) *An Essay on Man: An Introduction to a Philosophy of Human Culture*. New Haven and London: Yale University Press.

Castells, M. (1977) *The Urban Question: a Marxist Approach*. London: Edward Arnold.

Certeau de, M. (1988) *The Practice of Everyday Life*. Berkeley: University of California Press.

Chamberlain, M. (1989) *Growing up in Lambeth*. London: Virago.

Clark, D. (1996) *Urban World / Global City*. London: Routledge.

Crook, S. (1998) 'Minotaurs and other monsters: "everyday life" in recent social theory'. *Sociology*, 32(3): 523–40.

Dictionary of National Biography (1901) (ed. S. Lee) Supplement Vol. III. London: Smith, Elder and Co.

Dunlop, F. (1991) *Thinkers of Our Time: Scheler*. London: The Claridge Press.

Eade, J. (ed.) (1997) *Living the Global City: Globalization as Local Process*. London: Routledge.

Elias, N. and Scotson, J.L. (1965) *The Established and the Outsiders: A Sociological Enquiry into Community Problems*. London: Frank Cass & Co. Ltd.

Elias, N. and Scotson, J.L. (1993) *Etablierte und Außenseiter*. Frankfurt/M.: Suhrkamp.

Giddens, A. (1991) *The Constitution of Society: Outline of the Theory of Structuration*. Cambridge: Polity.

Giddens, A. (1993) *Modernity and Self-Identity: Self and Society in the Late Modern Age*. Cambridge: Polity.

Giddens, A. (1994) *The Consequences of Modernity*. Cambridge: Polity.

Glaser, B. and Strauss, A. (1979) *The Discovery of Grounded Theory*. New York: Aldine Publishers.

Goethe, J.W. (n.d.) *Criticisms, Reflections and Maxims*. London and Felling-on-Tyne: The Walter Scott Publishing Co. Ltd.

Goffman, E. (1966) *Behaviour in Public Places: Notes on the Social Organization of Gatherings*. New York: The Free Press.

Goffman, E. (1972) *Relations in Public: Microstudies of the Public Order*. Harmondsworth: Penguin.

Goffman, E. (1990) *The Presentation of Self in Everyday Life*. London: Penguin.

Gottmann, J. (1989) 'What are cities becoming the centres of ? Sorting out the possibilities', in R.V. Knight and G. Gappert (eds) *Cities in a Global Society*. London: Sage.

Gottmann, J. and Harper, R.A. (eds) (1990) *Since Megalopolis: The Urban Writings of Jean Gottmann*. Baltimore and London: The Johns Hopkins University Press.

Gower, G. (n.d.) *Streatham Common: Places of Historical Interest*. London: Local History Publications.

Gower, G. (1990) *A Brief History of Streatham*. London: Streatham Society.

Gower, G. (1993) *Sherlock Holmes in Streatham*. London: Local History Publications.

Grathoff, R. (1989) *Milieu und Lebenswelt: Einführung in die phänomenologische Soziologie und die sozialphänomenologische Forschung*. Frankfurt: Suhrkamp.

Grathoff, R. (1994) 'Von der Phänomenologie der Nachbarschaft zur Soziologie des

Nachbarn', in W. Sprondel (ed.) *Die Objektivität der Ordnungen und ihre Kommunikative Konstruktion: für Thomas Luckmann*. Frankfurt/M.: Suhrkamp.

Hall, J.M. (1990) *Metropolis Now: London and its Region*. Cambridge: Cambridge University Press.

Hannerz, U. (1980) *Exploring the City: Inquiries Towards an Urban Anthropology*. New York: Columbia University Press.

Hannerz, U. (1992) 'Cosmopolitans and locals in world culture', in M. Featherstone (ed.) *Global Culture: Nationalism, Globalization and Modernity*. London: Sage.

Hannerz, U. (1998) *Transnational Connections: Culture, People, Places*. London: Routledge.

Harvey, D. (1993) *The Condition of Postmodernity: An Enquiry into the Origins of Cultural Change*. Oxford: Blackwell.

Harvey, D. (1996) *Justice, Nature & the Geography of Difference*. Oxford: Blackwell.

Jackson, A.A. (1973) *Semi-Detached London: Suburban Development, Life and Transport, 1900–39*. London: Allen & Unwin.

Jessop, B. and Stones, R. (1992) 'Old city and new times: economic and political aspects of deregulation', in L. Budd and S. Whimster (eds) *Global Finance & Urban Living: A Study of Metropolitan Change*. London: Routledge.

Jones, E. (1990) *Metropolis*. Oxford and New York: Oxford University Press.

Kant, I. (1981) *Von den Träumen der Vernunft: Kleine Schriften zur Kunst, Philosophie, Geschicht und Politik*. Leipzig und Weimar: Kiepenheuer.

Keim, K.D. 'Milieu und Moderne. Zum Gebrauch und Gehalt eines nachtraditionalen sozial-räumlichen Milieubegriffs'. *Berliner Journal für Soziologie*, 7(3): 387–99.

King, A.D. (1991) *Global Cities: Post-Imperialism and the Internationalization of London*. London: Routledge.

King, A.D. (ed.) (1996) *Re-Presenting the City: Ethnicity, Capital and Culture in the 21st-Century Metropolis*. London: Macmillan.

Kiwitz, P. (1986) *Lebenswelt und Lebenskunst: Perspektiven einer kritischen Theorie des sozialen Handelns*. München: Wilhelm Fink.

Knight, R.V. and Gappert, G. (eds) (1989) *Global Cities in a Global Society*. London: Sage.

Kureishi, H. (1990) *The Buddha of Suburbia*. London and Boston: Faber and Faber.

Lambeth Unitary Development Plan (1992) London: Lambeth Borough Council. (On deposit.)

Lash, S. and Urry, J. (1994) *Economies of Signs and Space*. London: Sage.

Lefèbvre, H. (1997) *Writings on Cities* (edited by E. Kofman and E. Lebas). Oxford: Blackwell.

Lefèbvre, H. (1990) *Die Revolution der Städte*. Frankfurt/M.: Anton Hain.

Lindner, R. (1990) *Die Entdeckung der Stadtkultur: Soziologie aus der Erfahrung der Reportage*. Frankfurt/M.: Suhrkamp.

Loobey, P. and Brown, J. (1993) *Streatham in Old Photographs*. Phoenix Mill: Stroud.

Lutherans in London (1991a) *News from St Anne's Lutheran Church*. May.

Lutherans in London (1991b) *News from St Anne's Lutheran Church*. October.

McAuley, I. (1993) *Guide to Ethnic London*. London: Immel.

Maffesoli, M. (1996) *The Time of the Tribes: The Decline of Individualism in Mass Society*. London: Sage.

Melucci, A. (1996) *The Playing Self: Person and Meaning in a Planetary Society*. Cambridge: Cambridge University Press.

Merriman, N. (ed.) (1993) *The Peopling of London: Fifteen Thousand Years of Settlement from Overseas*. London: Museum of London.

Mingione, E. (1986) Urban sociology beyond the theoretical debate of the seventies. *International Sociology*, 1(2): 137–53.

Mumford, L. (1991) *The City in History: Its Origins, its Transformations, and its Prospects*. Harmondsworth: Penguin.

Myers, S.D. (1949) *London South of the River*. London: Paul Elek.

Nietzsche, F. (1979) *The Use and Abuse of History*. Indianapolis: The Liberal Arts Press.

Pahl, R.E. (1973) *Patterns of Urban Life*. London Longman.

Park, R.E. (1974) 'The mind of the hobo: reflections upon the relation between mentality and locomotion', in R.E. Park, W. Burgess and R.D. McKenzie (eds) *The City*. Chicago and London: University of Chicago Press.

Park, R.E., Burgess, W. and McKenzie, R.D. (eds) (1974) *The City*. Chicago and London: University of Chicago Press.

Plato (1977) *Phaedrus & Letters VII and VIII*. Penguin Classics. Harmondsworth: Penguin.

Raban, J. (1990) *Soft City*. London: Collons Harvill.

Rabinow, P. (1989) *French Modern: Norms and Forms of the Social Environment*. Cambridge, MA: MIT Press.

Robertson, R. (1993) *Globalization: Social Theory and Global Culture*. London: Sage.

Robertson, R. (1995) 'Glocalization: time–space and homogeneity–heterogeneity', in M. Featherstone, S. Lash and R. Robertson (eds) *Global Modernities*. London: Sage.

Reeves, G. (1986) *Palace of the People*. London: Bromley Library Service.

Ritzer, G. (1993) *The McDonaldization of Society: An Investigation into the Changing Character of Contemporary Social Life*. London: Pine Forge Press.

Sassen, S. (1991) *The Global City: New York, London, Tokyo*. Princeton: Princeton University Press.

Sassen, S. (1994) *Cities in a World Economy*. London and New Delhi: Pine Forge Press.

Savage, M. and Warde, A. (1996) 'Cities and uneven economic development', in R.T. LeGates and F. Stout (eds) *The City Reader*. London: Routledge.

Scheler, M. (1957) Gesammelte werke Bd 10 *Schriften aus dem Nachlaß: Zur Ethik und Erkenntnislehre*. Bern: Francke.

Scheler, M. (1973) *Formalism in Ethics and Non-Formal Values: A New Attempt toward the Foundation of an Ethical Personalism*. Evanston: Northwestern University Press.

Scheler, M. (1976) Gesammelte Werke Bd 9 *Späte Schriften*. Bern und München: Francke.

Schütz, A. (1966) *Collected Papers*. Vol. III 'Studies in Phenomenological Philosophy'. The Hague: Nijhoff.

Schütz, A. (1967) *Collected Papers*. Vol. I 'The Problem of Social Reality'. The Hague: Nijhoff.

Schütz, A. (1970) *Reflections on the Problem of Relevance* (edited by R.M. Zaner). New Haven and London: Yale University Press.

Schütz, A. (1971) *Collected Papers*. Vol. II 'Studies in Social Theory'. The Hague: Nijhoff.

Seaman, L. (1973) *Life in Victorian London*. London: B.T. Batsford.

Sennett, R. (1993) *The Fall of Public Man*. London: Faber and Faber.

Sexby's History of Streatham (1989) London: Local History Reprints.

Shakespeare, N. (1986) *Londoners*. London: Sidgwick & Jackson.

Silverman, D. (1989) *Qualitative Methodology & Sociology*. Aldershot: Gower.

Simmel, G. (1969) *The Sociology of Georg Simmel* (edited by K.H. Wolff). London, New York: Macmillan/The Free Press.

Smith, D. (1988) *The Chicago School: A Liberal Critique of Capitalism*. London: Macmillan.

Smith, D.A. (1996) *Third World Cities in Global Perspective: The Political Economy of Uneven Urbanization*. Oxford: Westview Press.

Spengler, O. (1971) *The Decline of the West*. London: Allen & Unwin.

Strauss, A. and Corbin, J. (1990) *The Basics of Qualitative Research: Grounded Theory Procedures and Techniques*. London: Sage.

Streatham Association (1993) *Annual Report*. London: Streatham Association.

Srubar, I. (1988) *Kosmion: Die Genese der pragmatischen Lebenswelttheorie von Alfred Schütz und ihr anthropologischer Hintergrund*. Frankfurt/M.:Suhrkamp.

Tames, R. (1992) *A Traveller's History of London*. Adlestrop: The Windrush Press.

Thorns, D.C. (1972) *Suburbia*. London: MacGibbon and Kee.

Tomlinson, J. (1994) 'A phenomenology of globalization? Giddens on global modernity'. *European Journal of Communication*, 9: 149–72.

Trevelyan, G.M. (1946) *English Social History: A Survey of Six Centuries*. London: Longmans, Green and Co.

Tull, G.F. (1974) *In the Willows: The Story of St Anne and St Agnes*. Ashford: The Manor Press.

Vaitkus, S. (1991) *How is Society Possible? Intersubjectivity and the Fiduciary Attitude as a Problem of the Social Group in Mead, Gurwitsch and Schütz*. Dordrecht: Kluwer.

Vulliamy, C.E. (1936) *Mrs Thrale of Streatham*. London: Jonathan Lape.

Waldenfels, B. (1985) *In den Netzen der Lebenswelt*. Frankfurt/M.: Suhrkamp.

Weber, M. (1966) *The City*. New York: The Free Pres.

Weinreb, B. and Hibbert, C. (eds) (1983) *The London Encyclopedia*. London: Macmillan.

Willows, H. (ed.) (1988) *A Guide to Worship in Central London*. London: Fowler Wright Books.

Wirth, L. (1969) *On Cities and Social Life*. London and Chicago: University of Chicago Press.

Young, M. and Willmott, P. (1962) *Family and Kinship in East London*. Harmondsworth: Penguin.

INDEX

affective field 74, 76, 81; stable 70–2; unsettled 72

Albrow, M. 8, 11, 16, 22, 104, 142, 145, 165, 166; *et al* 18, 21

Anderson, Ulla 122; affective field of 70–2, 76; as settled cosmopolitan 28–30

Appadurai, A. 9, 10, 13, 14, 17, 18, 72, 134, 135, 136, 139, 145, 171, 177

appresentational situations 133, 134–5, 136, 139

Aristotle 5, 116, 117

Arnold, F. 99, 109

Augé, M. 66

back regions 62–3

Bailey, P. 164

Beck, U. 7, 26, 48, 161

Berger, P.L. *et al* 9, 22, 23, 58, 59, 172

biographical approach: case histories 27–41; choice of 25–6; and concept of milieu 26–7; critics of 26

Borer, M.C. 100, 107

Boswell, James 103

Bourdieu, P. 26

Braithwaite, Ira: detachment from neighbours 156–7; leaving the estate/finding herself 48–54; and meeting strangers 149–52; as the metropolitan 35–7; sense of home 80

Brandon, P. 107–8

Braudel, 7

Brenner, N. 12

Bromhead, H.W. 98, 103, 108, 109, 110

Budd, L. and Whimster, S. 12, 13

Burke, Edmund 103

Cadman, D. and Payne, G. 11

Calacuri, Nicos: as local entrepreneur 32–4; and meeting unknown strangers 147–8; sociability/ uncertainty 144–5; symbolic attachment to place 72

Cassirer, E. 18, 20, 120–2, 175, 176

Castells, M. 8

Certeau, M. de 74, 75, 112, 116–18, 176

Chamberlain, M. 163–5

Chicago School 7

city: ambiguity of 127; and capitalism 8; as centre of gravity 7–8, 11; ethnoscape 135–6; flexible concept of 122; as global 12–15; and globalization 8–12; heterogenetic 153; medieval 6; melodrama of 42; and modern civilization 5–8; modernist 6–7; of neighbours 153–7; as place of encounters 42; as place of self-actualization 128; 'seeing' of 122; size of 5; social order in 5–6; sociological 7; technoscape of 135–6; understanding 4–5; as world city 6

City of London 100, 107

Clark, D. 11

community: experience 82; neighbourhood 161–71; street village 163–4

concept city: described 116–17; and symbolic space 117–23

cosmion: defined 133; described 133–40; as intersection 150
cosmopolitanism 149–50, 152
Crook, S. 175
culture 127–8, 174

disembedding 23–4, 43, 60, 77, 81, 172; relevances of action 64–5; trust/care relationships 160–1
Doyle, A.C. 113
DuCane, Peter 99
DuCane, Richard 99
Dunlop, F. 82

Eade, J. 12, 172
East India Company 99, 100
Elias, N. and Scotson, J.L. 89, 176
environment 47–9, 119, 131; of like-mindedness 81–90
everyday life 129–31, 132–3
extended milieu 18–22, 23, 142–3; lived reality of 151, 172–4; miniature of 125, 173; phenomenology of 18–22; situated 61–2, 63, 68, 73–4, 76–7, 80, 89–90; *see also* milieu

field of action 74; decentredness of 69–70
flâneur 120, 124, 150
Ford, Harold 122, 138; attitude to crowds 148–9; character of 63; experience of everyday life 129; as global business man 30–2; and his neighbours 154–5, 169–70; like-mindedness of 88–90; and locale 166–7; and meeting unknown strangers 148–9; mobility of 61–8; and real experience/interest 130, 131; travel sense 63–4

Giddens, A. 9, 22–4, 43, 59, 60, 70, 77, 81, 101, 109, 158–9, 161
Glaser, B. and Strauss, A. 26
global city 10–12, 172–4; biographies of lives in 25–41; coexistence of social/cultural worlds in 14–15; as cosmion 133–40; detachment in 153; excessive stimulation in 127–9, 135–7; and its microglobalized hinterland 12–15; socioscape of 161–71

global cultural economy 9–10, 139; *en miniature* 173
globalization 82, 142, 173; and phenomenology 16–18; vs urbanization 8–12
globalized world city 136; described 13, 14–15; London as 133; as prism of global cultural economy 132, 173–4
Goethe, J.W. 13, 91
Goffman, E. 18, 19–20, 24, 26, 62–3, 69, 81, 157, 160, 168, 175
Gottman, J. 13, 60; and Harper, R.A. 9
Gower, G. 94, 95, 98, 99–100, 103, 104, 105, 106, 108, 109, 110
Grathoff, R. 19, 26, 153, 158
Grauer, Rolf: and his neighbours 161, 162–3; pursuit of music/singing 128–9; as reluctant explorer 39–41; sexual coming out 43–8

Hankey, John 99
Hannerz, U. 65, 70, 83, 85, 152, 175
Harvey, D. 11, 12, 14, 60, 75, 82, 91–2, 99, 101, 108–9, 111, 114, 116, 127, 134, 141
history 112, 113–14
Holmes, Sherlock 113
home: Johnsonian 104; as significant place 77–81
Howland family 100, 101
Husserl, E. 16

in-group/immigrant 144, 146–7
Industrial Revolution 100, 101–2

Jackson, A.A. 108
Jessop, B. and Stones, R. 11
Johnson, Dr Samuel 103–4, 132, 165
Jones, E. 7

Kant, I. 157
King, A.D. 10–11, 12, 14, 56, 93, 107
Kiwitz, P. 42, 116
Keim, K.D. 21
Knight, R.V. and Gappert, G. 8, 12
Kureishi, H. 12

Lash, S. and Urry, J. 9, 13, 17, 60, 61, 136, 140
Lefèbvre, H. 9
like-mindedness: concept of 82; and globalization 82–3; non-localized social 84, 87–90
Lindner, R. 7
locale: attachment to 89–90; attitude to 166–7; convergence with milieu 77; defined 58; delinking of 74–7, 84–90; as home 78–81; transcendence of 111–12
London Government Act (1899) 108
Loobey, P. and Brown, J. 101

Maffesoli, M. 82
Massingbird, John 99
mediascapes 72, 137, 140
Melucci, A. 7, 18, 26
Merriman, N. 14, 25, 56, 101, 132
metropolis: as centre of gravity 7–8; Greek concept 5; and modernity 6–7; and social order 5–6; spatial structuration of 7; as world city 6
microglobalization 12–15, 76, 115, 132; as threat to local milieu 58
milieu 42–3; care-taker function 159–60; concept of 26–7; convergent 74–7; decentred 68–9; defined 58–9; delinking of 23, 74–7, 84–90, 104–5; disrupting local 54–9; of local tradition 112; mobile/generalized 61–8; non-localized social 83–4; as personal orientation in the world 118–19; poly-centred 73–4; practical 119–20; re-establishing 123; separate 154; stratification of 105, 106; structural/practical difference 64; touching/intersecting of 168–9; transcendence of immediate locale 111–12; as *Umwelt* 19–20; uprooting of 43–54, 123; as value-related environment 47, 69; *see also* extended milieu
Mingione, E. 8, 9
mobility 61–8; emptied out 62; from place to place 77; permanent/unpredictable 65–6; routines associated with 65

Mumford, L. 5–6, 8–9, 42, 116, 133–4, 174
Myers, S.D. 103, 105

nearest other 161, 174; detached 153, 156, 157–8; distant 158–60, 163; as gate-keeper 158, 159; generalized 156
neighbourhood community 161–71; as close/shared 164; confined 165; retired/professional 167–8; rules/standards in 164; sociability of 168–71; street village 163–4; transformation in 163–4, 166, 168, 169–71
neighbours 153–7; approved detachment from 155–7, 163; as detached 157, 165; distant 157–61; and maintenance of respective optional fields 154; as nearest other 153, 156, 157–8, 159, 161, 165; nodding acquaintance of 168–9; relationship with 154; symbolic order of 154–6; trust/care relationship with 157, 159–61; turnover of 162
Nietzsche, F. 114

other *see* stranger/other

Pahl, R.E. 100, 104, 106, 111, 164
Park Hill house 105
Park, R.E. 7, 61
Penhaligan, Sarah: and attitude to locale 166, 167–8; and her neighbours 160, 170–1; link with locale/home 78–80; management of space 123–4; as mobile cosmopolitan 27–8; and negotiation of city 152; work/leisure friends 75–6, 128
phenomenology: of everyday life 16–18; of extended milieux 18–22; re-locating the concept 22–4
place: phantasmagoric 101; relational aspect 98; symbolic attachment to 72–3; *see also* significant place
Plato 4
practiced city 117–28
Pratt, George 105–6
Pratt's department store 105–6, 110, 111–12

Raban, J. 42, 115, 117, 125
Rabinow, P. 77
re-embedding 81–90, 172
Reading, Herbert 162; and attitude to locale 166; and concern for milieu 146–7; lived reality of 144; and maintaining/regaining local milieu 54–9; as retired local 34–5; and street village community 163–4
Reeves, G. 109
relevances of action: decontextualized 66–7; disembedded 64–5
Reynolds, Joshua 103
Ritzer, G. 66
Robertson, R. 23, 76, 80, 81
routine activities 66

St Anne and St Agnes Church 84–90, 119, 143
St Leonard's church 99, 104
Sassen, S. 10, 12
Savage, M. and Warde, A. 9
Scheler, M. 18, 19, 47, 48, 53, 59, 82, 112, 118–19, 130
Schütz, A. 15, 18, 19, 43, 64, 66, 67, 68, 74, 87, 118, 119, 129–30, 133, 134, 136, 139, 143, 167, 176, 177
Sennett, R. 140, 141–2, 149
Shakespeare, N. 25
significant place 68, 77, 89; and affective field 69–70, 72, 73–4; home as 77–81; see also place
Silverman, D. 93
Simmel, G. 6–7, 127–9, 141, 169–70
Smith, D.A. 7, 9
sociability 125–6, 143, 144–5, 167–71
socioscape 149, 166–7, 173
sociosphere 140–53; coexisting 150–2; defined 142–3; globalized 150; and maintenance of extended milieux 143–4; non-localized 163; realities of 145–8; retired elderly/busy professional 167–8; single local 165; socialising in 168–71; stratification of 165; and transformation of in-group/stranger 146–7
Socrates 4–5
Soft City 115, 116–17

space: active/symbolic 121; disciplinary vs lived 117; intuitive 121; significant 70; territorial/geographical 70; see also time–space compression, symbolic space
Spengler, O. 6
Srubar, I. 139, 176, 177
stranger/other: accommodation of 150–2; concept of 141; crowds of 148–9; lived experience of 142–5; meeting/encountering 147–50; as outsider 141; relationship of 140–1; technical 152–3; as threat 150; transformation of group 147–8; as unknown 141–2, 143, 147–8
Streatham 139, 165, 167; absorbed by metropolis 106–8; changes in built environmen 109–10; closure of Pratt's in 97–8; decline 96; demographics 93–4; Dr Johnson's 101–6; environment 94–5; High Road 98–101; history/development 98–9; image/reputation 95–6, 110, 111; location 93; struggle for local identity/disruption of local milieu 108–14; suburbanization of 107–8, 109, 110; would-be West End of London 93–8
Streatham Association 95, 98
Streatham Place house 103–4
Streatham Society 95
suburbs 107–8, 109, 110
symbolic discourse 135–8, 139–40
symbolic disorientation 120
symbolic space 117–21, 124, 125; generation of 121–3; graffito example 134–5; see also space
symbolic territory of the self 69, 81

Tames, R. 13
Tate, Henry 105
third territory 125–6, 168
Thorns, D.C. 107
Thrale family 103, 105, 165
Tilney, Edmund 99
time–space compression 75–7, 82, 91, 92–3, 108, 109, 115, 132, 134, 142, 145, 168, 172; see also space

transport 94–5, 97, 106, 107, 109, 125–6
Trevelyan, G.M. 98, 101–3
Tull, G.F. 84

uprooting 123; external 43–8; internal
    48–54

Velde, Barbara van der 137, 143;
    decentredness of 69; and distant
    neighbour 158, 159; as expatriate 37–9;
    interest in whole city 118, 119–20,
    122, 124; like-mindedness of 88–90;
    management of symbolic space 123,
124; and sense of home 73–4
Vulliamy, C.E. 102, 103

Waldenfels, B. 68, 69–70, 79, 80, 125,
    153
Weber, M. 6
Weinreb, B. and Hibbert, C. 99, 103,
    105, 110
Willows, H. 84, 176
Wirth, L. 7, 134, 173
world city 3, 12, 13, 14, 172–4

Young, M. and Willmott, P. 55, 163